# 2nd International Conference on Power Electronics and Advanced Control (PEAC 2022)

AA002460

Journal of Physics: Conference Series Volume 2434

Guilin, China
18-20 November 2022

ISBN: 978-1-7138-6775-3
ISSN: 1742-6588

**Printed from e-media with permission by:**

Curran Associates, Inc.
57 Morehouse Lane
Red Hook, NY 12571

**Some format issues inherent in the e-media version may also appear in this print version.**

This work is licensed under a Creative Commons Attribution 3.0 International Licence.
Licence details: http://creativecommons.org/licenses/by/3.0/.

No changes have been made to the content of these proceedings. There may be changes to pagination and minor adjustments for aesthetics.

Printed with permission by Curran Associates, Inc. (2026)

For permission requests, please contact the Institute of Physics
at the address below.

Institute of Physics
Dirac House, Temple Back
Bristol BS1 6BE UK

Phone:   44 1 17 929 7481
Fax:      44 1 17 920 0979

techtracking@iop.org

**Additional copies of this publication are available from:**

Curran Associates, Inc.
57 Morehouse Lane
Red Hook, NY 12571 USA
Phone: 845-758-0400
Fax:    845-758-2633
Email:  curran@proceedings.com
Web:   www.proceedings.com

# TABLE OF CONTENTS

Preface

Peer Review Statement

A Transparent Tunable Broadband Microwave Absorber Based on Multi-Layer Structure by
Patterned Graphene ................................................................................................................. 1
   *Limin Ma, Zhenghua Wang, Xiaoyu Han*

Robust Planning Method for Photovoltaic Microgrid Energy Storage Considering Source-Load
Flexibility Resources ............................................................................................................... 6
   *Xianghai Xu, Jiayi Shang, Zhiyuan Chen, Zhipeng Zhang*

Design and Research of the Intelligent Water Temperature Monitoring System .............................. 12
   *Zhuolin Wang*

Harmonic Voltage Amplitude Calculation and Fundamental Amplitude Correction Based on
Synchronized SVPWM of Two-Level VSI Under Low Switching Frequency ................................... 20
   *Yueqiang Hu, Fengguang Jiang*

On-Line Monitoring of Instantaneous Over-Voltage Intelligent Monitoring of EMUs ...................... 26
   *Zhongjiang Sun, Suyue Liu, Zhiming Yang, Shuangfeng Xu*

Rational Design of Future Potential Electric Aircraft ................................................................. 36
   *Mingze Xi*

Adaptive Neural Control for Chaotic Permanent Magnet Synchronous Motor with Asymmetric
Input Saturation ..................................................................................................................... 41
   *Hongshan Liu, Huibo Liu, Yanwei Zhao*

**Author Index**

# Preface

The International Conference on Power Electronics and Advanced Control is a leading international conference. It aims at building an international platform for the communication and academic exchange among participants from various fields related to power electronics and advanced control. Here, scholars, experts, and researchers are welcomed and encouraged to share their research results and inspirations. It is a great opportunity to promote academic communication and collaboration worldwide.

This volume contains the papers presented at the 2nd International Conference on Power Electronics and Advanced Control (PEAC 2022), held during November 11st-13rd, 2022 in Guilin, China (virtual event). For the safety concern of all participants, we decided to hold it as a virtual conference which is also effective and convenient for academic exchange and communication. Everyone interested in this field were welcomed to join the online conference and to give comments and raise questions to the speeches and presentations.

The online conference was composed of keynote speeches, oral presentations, and academic investigation, attracting about 50 individuals from all over the world. We have invited three sophisticated professors to perform keynote speeches. Among them, Prof. Sudip K. Mazumder from University of Illinois, USA is one of our keynote speakers and delivered a dramatic speech. He has around 30 years of professional experience, held research and development and design positions in leading industrial organizations, served as a Technical Consultant for several industries, and has been serving as a Regional Distinguished Lecturer for the IEEE Power Electronics Society for the US region since 2021. Besides, he has published more than 250 refereed papers, delivered over 110 keynote/plenary/distinguished/invited presentations, and received 55 sponsored research studies since joining the UIC. All the keynote speakers made brilliant speeches and shared unique experience and insights, and we are hoping the pandemic soon come to an end and we could see each other face to face next year.

PEAC 2022 received a great many submissions in the areas of power electronics and advanced control. Each submission was reviewed by at least two expert reviewers and the committee picked out some excellent papers that are included in the proceedings, including but not limited to the following topics: Modeling and Simulation, Grid Monitoring Technology, Distributed Control Systems, Roadside Equipment Systems, etc.

On behalf of the Conference Organizing Committee, we would like to thank the Technical Program Committee members and external reviewers for their hard work in reviewing and selecting papers. And we would like to acknowledge all of those who supported PEAC 2022. In particular, our special thanks go to the Journal of Physics: Conference Series. Hopefully, all participants and other interested readers can benefit significantly and scientifically from the proceedings and find it rewarding in the process.

The Committee of PEAC 2022

# Committee member

**Committee Chair**
Prof. Chris Mi, San Diego State University, USA

**Publication Chair**
Prof. Sudip K. Mazumder, University of Illinois, USA

**Technical Program Committee Chair**
Prof. Brij B. Gupta, Indian Institute of Technology, India

**Technical Program Committee**
Prof. Ning Sun, Nankai University, China
Prof. Dr. Murat Tolga OZKAN, Gazi University, Turkey
Researcher Eric Thompson Brantson, University of Mines and Technology, Ghana
Assoc. Prof. Wei Wei, Xi'an University of Technology, China
Prof. Pradip Jawandhiya, Pankaj Laddhad Institute of Technology and Management Studies Buldana, India
Prof. Noreddine Gherabi, Computer Science at Sultan Moulay Slimane University, Morocco
Assoc. Prof. Ahmed A. Abd El-Latif, Menoufia University, Saudi Arabia
Assoc. Prof. Srimanti Roychoudhury, Daffodil International University, Bangladesh
Prof. Predrag Ivanis, University of Belgrade, School of Electrical Engineering, Serbia
Assoc. Prof. Prashant Mani, SRM University, India
Assoc. Prof. Ahmed Okasha, Cairo University, Egypt
Assoc. Prof. Rabie Ramadan, Cairo University, Egypt

**Organizing Committees**
Prof. Arindam Ghosh, Curtin University, Australia
Prof. Minoru Sasaki. Gifu University, Japan
Prof. Ramash Kumar K, University of Madras, Inida
Prof. Ben cheikh Hamida, University Amar Telidji Laghouat Algeria, Algeria
Assoc. Prof. Raghvendra Chaudhary, Department of Electrical Engineering, Indian Institute of Technology Kanpur, Inida
Assoc. Prof. Thangaprakash Sengodan, Anna University, India
Assoc. Prof. Sergio Vázquez, University of Seville, Spain
Assoc. Prof. Muhyiddine Jradi, University of Southern Denmark, Denmark

# Peer Review Statement

All papers published in this volume have been reviewed through processes administered by the Editors. Reviews were conducted by expert referees to the professional and scientific standards expected of a proceedings journal published by IOP Publishing Publishing.

- **Type of peer review:** Single Anonymous
- **Conference submission management system:** Morressier
- **Number of submissions received:** 31
- **Number of submissions sent for review:** 28
- **Number of submissions accepted:** 13
- **Acceptance Rate (Submissions Accepted / Submissions Received × 100):** 41.9
- **Average number of reviews per paper:** 2
- **Total number of reviewers involved:** 4
- **Contact person for queries:**
  **Name:** Xuexia Ye
  **Email:** xx.ye@keoaeic.org
  **Affiliation:** AEIC Academic Exchange Information Centre

# A Transparent Tunable Broadband Microwave Absorber Based on Multi-layer Structure by Patterned Graphene

**Limin Ma, Zhenghua Wang and Xiaoyu Han**

College of Automation Engineering Nanjing University of Aeronautics and Astronautics, Nanjing, China;

liminma@nuaa.edu.cn

**Abstract:** As the increasingly complicated electromagnetic environment, it is crucial to develop the performance of transparent microwave absorbers. Here we proposed a transparent tunable broadband microwave absorber, which is consisted of double-layer graphene rings and metallic grids. The microwave shielding properties can be changed by adjusting the dielectric layer thickness, graphene Fermi energy and relaxation time. Based on the simulated results, the proposed transparent microwave absorber can achieve microwave absorption of over 85% at a frequency range of 4.61–9.75 GHz, indicating a relative absorption bandwidth nearly 71.6% and a normalized transmittance of 88.8% in the visible band.

## 1. Introduction

The electromagnetic environment is becoming increasingly complicated [1]. To avoid electromagnetic interference (EMI), EMI shielding emerged and attracted extensive attention worldwide. The application of EMI shielding for optically transparent objectives is one of the difficulties and hotspots due to the conflict between strong EMI shielding performance and high optical transparency [2].

Transparent conductive coatings exhibit optical transparency and strong electromagnetic reflection. For instance, metallic grids can achieve visible transmittance over 90% and excellent EMI shielding efficiency, simultaneously [3]. In 2019, Wang et al. proposed an electromagnetic Ag shielding films that realized an 26 dB electromagnetic interference SE and 96.5% visible transmittance.

Graphene, as the first available two-dimensional material, possesses many unique properties, including good conductivity, ultra-thin structure, excellent thermal stability, broadband optical transparency, and microwave absorption. The above unique properties make graphene one of the potential paves to achieve transparent microwave absorption [4]. Wu et al. produced a microwave absorber with multi-layer graphene and a metal plate that achieved broadband absorption. The relative absorption bandwidth is 28% at the frequency range of 125–165 GHz. In 2022, we proposed a transparent wideband microwave absorber composed of metal rings and graphene. Over 90% absorption in the 25.2-35.2 Ghz frequency span with 10GHz bandwidth. It yield broadband absorption characteristics from visible to infrared light and has high compatibility [6].

Herein, to further expand the relative absorption bandwidth of the transparent microwave absorbers, we proposed a tunable broadband absorber that is consisted of patterned graphene layers, dielectric layers, and metallic plates. The patterned graphene layers are formed by periodically arranged rings, and the transparent metallic grids act as the metallic plate. Results indicate the absorber has a relative absorption bandwidth of 71.6% at a frequency range of 4.61–9.75 GHz and visible transmittance of 88.8% simultaneously.

Content from this work may be used under the terms of the Creative Commons Attribution 3.0 licence. Any further distribution of this work must maintain attribution to the author(s) and the title of the work, journal citation and DOI.

Published under licence by IOP Publishing Ltd

## 2. Results and of Discussions

Figure 1 shows the proposed transparent microwave absorber. It consists of two graphene layers and a copper (Cu) layer, and each layer is separated by dielectric spaces. The top layer consists of an array of graphene rings with an inside radius $r_1 = 0.16$ mm and an outside radius $r_2 = 0.92$ mm, as shown in Figure 1c. The middle layer consists of an array of graphene rings with an inside radius $r_3 = 0.16$ mm and an outside radius $r_4 = 0.8$ mm as shown in Figure 1d, while the dielectric space between two graphene layers has a permittivity of $\varepsilon = 3.5$ with a thickness of $t_1 = 2.2$ mm. The bottom layer consists of an array of copper rings with an inside radius $r_5 = 0.99$ mm and an outside radius $r_6 = 1$ mm as shown in Figure 1e, while the dielectric space between the middle graphene layer and the copper rings layer as a permittivity of $\varepsilon = 3.5$ with a thickness of $t_2 = 3$ mm. The bottom copper conductivity $\sigma = 5.71 \times 10^7$ Sm$^{-1}$.

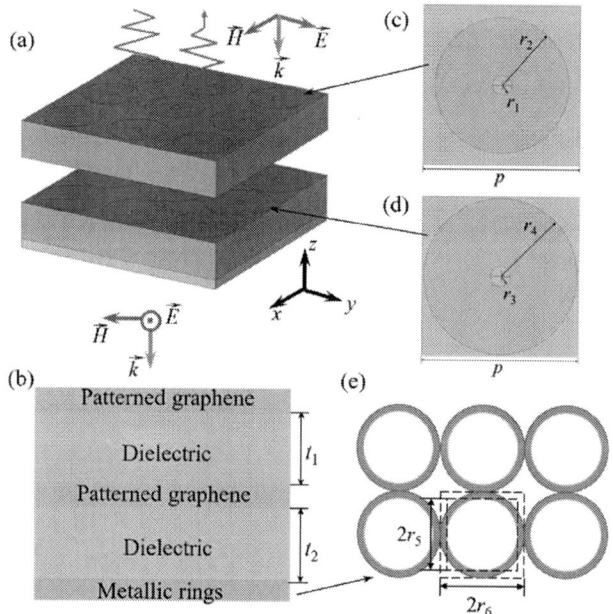

Figure 1. (a) The perspective view of the broadband microwave absorber. (b) The side view of the absorber. (c) The unit cell of the top. (d) middle. (e) bottom layer.

The conductivity of graphene can be expressed as

$$\sigma\left(\omega,\Gamma,\mu_c,T\right) = \sigma_{\text{intra}}\left(\omega,\Gamma,\mu_c,T\right) + \sigma_{\text{inter}}\left(\omega,\Gamma,\mu_c,T\right) \tag{1}$$

where $\sigma_{\text{intra}}$ and $\sigma_{\text{inter}}$ are the conductivity of the intra-band and inter-band, respectively, which can be expressed as

$$\sigma_{\text{intra}}\left(\omega,\Gamma,\mu_c,T\right) = \frac{-ie^2}{\pi\hbar^2(\omega - j2\Gamma)} \int_0^\infty \varepsilon\left(\frac{\delta f_d(\varepsilon)}{\delta\varepsilon} - \frac{\delta f_d(-\varepsilon)}{\delta\varepsilon}\right) d\varepsilon \tag{2}$$

$$\sigma_{\text{inter}}\left(\omega,\Gamma,\mu_c,T\right) = \frac{ie^2(\omega - j2\Gamma)}{\pi\hbar^2} \int_0^\infty \frac{f_d(-\varepsilon) - f_d(\varepsilon)}{(\omega - i2\Gamma)^2 - 4(\varepsilon/\hbar)^2} d\varepsilon \tag{3}$$

$$f_d(\varepsilon) = \left(1 + \exp\left[\frac{\varepsilon - \mu_c}{K_B T}\right]\right)^{-1} \tag{4}$$

where $k_B$ is the Boltzmann constant, $\hbar$ is the Dirac constant, $e$ is the charge of an electron, $f_d(\varepsilon)$ is the Fermi-Dirac distribution, $\mu_c$ is the Fermi energy of graphene, $T$ is the Kelvin temperature, $\Gamma$ is the carrier scattering rate, and $\Gamma = 1/\tau$, where $\tau$ is the relaxation time of graphene. $\sigma_{intra}$ and $\sigma_{inter}$ can be further derived as

$$\sigma_{intra}\left(\omega, \Gamma, \mu_c, T\right) = -i\frac{e^2 K_B T}{\pi \hbar^2 (\omega - j2\Gamma)}\left[\frac{\mu_c}{K_B T} + 2\ln\left(e^{-\frac{\mu_c}{k_B T}} + 1\right)\right] \quad (5)$$

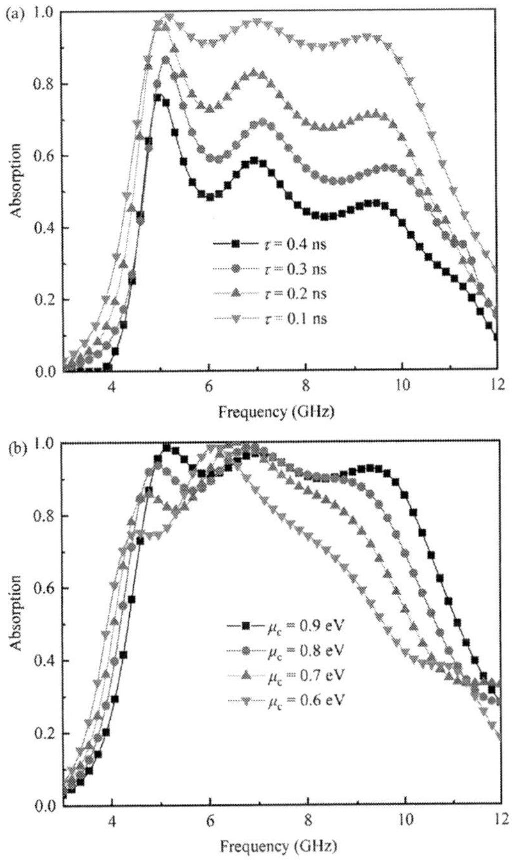

Figure 2. The absorption results as a function of frequency under different (a) the relaxation time of graphene $\tau$, and (b) the Fermi energy of graphene $\mu_c$.

$$\sigma_{inter}\left(\omega, \Gamma, \mu_c, T\right) = \frac{-ie^2}{4\pi\hbar}\ln\left(\frac{2|\mu_c| - (\omega - j2\Gamma)\hbar}{2|\mu_c| + (\omega - j2\Gamma)\hbar}\right) \quad (6)$$

The conductivity of graphene is decided by the Fermi energy $\mu_c$ and the carrier scattering rate based on Equations (5) and (6). Furthermore, the conductivity is related to the EMI shielding performance of graphene. Hence, we first analyze the influence of the Fermi energy $\mu_c$ and the carrier scattering rate $\Gamma$ on the EMI shielding performance of the proposed optically transparent microwave absorber. The commercial electromagnetic solver is used to analyze the suggested design. In simulated course, by applying periodic boundary conditions, using CST frequency-domain solver to analyze the absorber structure. The absorption $A$ through the suggested optically transparent microwave absorber can be determined by calculating the reflection coefficient ($S_{11}$) and transmission coefficient ($S_{21}$) as

$$A = 1 - R - T \tag{7}$$

$$R = \left|S_{11}\right|^2 \tag{8}$$

$$T = \left|S_{21}\right|^2 \tag{9}$$

The conductivity of graphene could be tuned via changing by chemical doping or applied bias voltage thus the electromagnetic waves transmission characteristic can be adjusted and tunable microwave absorption can be achieved [7-9]. Figure 2 shows the relationship between microwave absorption and the conductivity of graphene. Figure 2a shows that microwave absorption increases at the frequency range of 3–12 GHz with the decrease of the relaxation time of graphene $\tau$. The absorption increased to 97.18% from 70.82% at 5.07 GHz when the relaxation time of graphene $\tau$ decreased from 0.4 ns to 0.1 ns. As shown in Figure 2b, the bandwidth of microwave absorption broadens with increasing the Fermi energy of graphene $\mu_c$. The bandwidth of absorption over 85% increases from 1.58 to 5.25 GHz when the Fermi energy of graphene $\mu_c$ changes from 0.6 to 0.9 eV. Based on the above analysis, we set $\mu_c = 0.9$ eV, $\tau = 0.1$ ns.

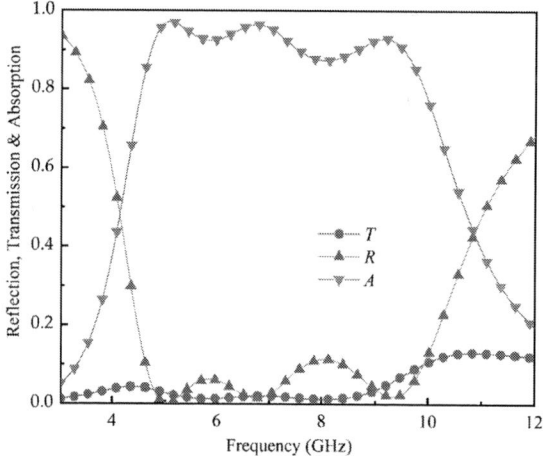

Figure 3. The calculated results of reflection, transmission, and absorption of the proposed optically transparent microwave absorber at 3–12 GHz.

As Figure 3 exhibits, the maximum absorption is 97.18% at 5.07 GHz of the absorber. In the range of 4.61 GHz to 9.75GHz, the absorption reaches more than 85% with a bandwidth of 5.14GHz. The relative absorption bandwidth $BW$ can be written as

$$BW = \frac{2 \times (f_H - f_L)}{f_H + f_L} \times 100\% \tag{10}$$

where is the upper limiting frequency of 85% absorption, $f_L$ is the lower limiting frequency of 85% absorption. The relative absorption bandwidth of the proposed optically transparent microwave absorber is 71.6%, as calculated by Equation (10).

Next, we discuss the normalized optical transmittance of the proposed microwave absorber. The proposed microwave absorber consists of two graphene ring layers and a metallic ring layer. The monolayer graphene optical transmittance is $T_{mg} = 97.7\%$ [10]. The surface resistance of mono-layer graphene is 327 $\Omega$ sq$^{-1}$ ($\mu_c = 0.3$ eV, $\Gamma = 3.80$ meV). The conductivity of graphene used here can be derived by Equation (6). Multilayer graphene could be taken as the parallel connection of monolayer graphene thus we can conclude that the surface resistance of 95 $\Omega$ sq$^{-1}$ can be achieved by four-layer graphene. The optical transmittance of top layer $T_{top}$, middle layer $T_{middle}$ and bottom layer $T_{bottom}$ can be calculated by

$$T_{top} = 1 - \frac{\pi\left(r_2^2 - r_1^2\right)}{p^2} \times \left(1 - T_{mg}^4\right) \tag{11}$$

$$T_{middle} = 1 - \frac{\pi\left(r_4^2 - r_3^2\right)}{p^2} \times \left(1 - T_{mg}^4\right) \tag{12}$$

$$T_{bottom} = 1 - \frac{\pi\left(r_6^2 - r_5^2\right)}{4r_6^2} \tag{13}$$

The optical transmittance of the proposed microwave absorber can be calculated by

$$T = T_{top} \times T_{middle} \times T_{bottom} \tag{14}$$

Based on the above analysis, the optical transmittance of the proposed microwave absorber is 88.8%.

## 3. Conclusions

From the above discussions, based on theoretical research and simulation results, the absorber composed of graphene and metal rings, which would yield tunable wideband absorption characteristics and high optical transparency compatibility. The final results indicate over 85% absorption in the 4.61-9.75GHz frequency span,the absorber accomplish maximum absorption of nearly 98% and normalized transmittance of 88.8% in the visible band.

## Acknowledgment

The authors acknowledge financial support from the Natural Science Foundation of Jiangsu Province (No. BK20190405).

## References

[1] Lu Z, Zhang Y, Lu X, Wang H and Tan J 2021 Two-step randomized design of multi-rings metallic mesh for ultra-uniform diffraction distribution Opt. Laser Technol. vol 144 pp 107396

[2] Han Y, Liu Y, Han L, Lin J and Jin P 2017 High-performance hierarchical graphene/metal-mesh film for optically transparent electromagnetic interference shielding Carbon (New York) vol 115 pp 34-42

[3] Wang G, Hao L, Zhang X, Tan S, Zhou M, Gu W, and Ji G 2022 Flexible and transparent silver nanowires/biopolymer film for high-efficient electromagneticinterference shielding J. Colloid Interface Sci. vol 607 pp 89-99

[4] Yi D, Wei X, and Xu Y 2017 Transparent Microwave Absorber Based on Patterned Graphene: Design, Measurement, and Enhancement IEEE Trans. Nanotechnol. vol 16 pp 484-490

[5] Li X, Feng G, and Lin S 2021 Ultra-wideband terahertz absorber based on graphene modulation Applied Optics vol 60 pp 3170-3175

[6] Ma L, Xu H, Lu Z, and Tan J 2022 Optically Transparent Broadband Microwave Absorber by Graphene and Metallic Rings ACS Appl. Mater. Interface vol 14 pp 17727-17738

[7] Gu M, Xiao B, and Xiao S 2018 Tunable THz perfect absorber with two absorption peaks based on graphene microribbons Micro Nano Lett. vol 13 pp 631-635

[8] Wu B, Yang Y, Li H, Zhao Y, Fan C and Lu W 2020 Low-Loss Dual-Polarized Frequency-Selective Rasorber With Graphene-Based Planar Resistor IEEE Trans. Antennas Propag. vol 68 pp 7439-7446

[9] Xu B, Gu C, Li Z, and Niu Z 2013 A novel structure for tunable terahertz absorber based on graphene Opt. Express vol 21 pp 23803-23811

[10] Geim A 2009 Graphene: Status and Prospects Science vol 324 pp 1530-1534

# Robust Planning Method for Photovoltaic Microgrid Energy Storage Considering Source-Load Flexibility Resources

**Xianghai Xu, Jiayi Shang, Zhiyuan Chen\*, Zhipeng Zhang**

Hangzhou Power Supply Company, State Grid Zhejiang Electric Power Co., Ltd., Hangzhou 310000, China

*Corresponding author's e-mail: chenzy_hzgd@126.com

**Abstract.** The microgrid based on distributed generation is one of the new forms of power system distribution network, and energy storage can provide important support for the access of distributed generation. Due to the shortcomings of the traditional photovoltaic microgrid energy storage method, the energy storage capacity is low. To improve the energy storage level of the photovoltaic microgrid, the robustness planning method of photovoltaic microgrid energy storage considering the flexibility resources of source, network and load is studied. Based on the flexibility resources of the microgrid, the balance index of flexibility supply and demand and the planning index of flexibility response capability is proposed. According to the source-network load algorithm, the optimal scheme of resource complementation is obtained. Based on the storage battery and supercapacitor of the photovoltaic microgrid, the photovoltaic microgrid energy storage robust configuration model is constructed. Comparing the energy storage planning method designed in this paper with two groups of traditional methods, the experimental results show that in the same energy storage time, the energy storage capacity of this method accounts for 50.49%, while that of the traditional group 1 and group 2 is 32.52% and 41.26%, respectively. The proposed PV microgrid robust planning method considering source-load flexibility is reasonable and effective in the energy storage resource allocation scheme, which is of great significance for promoting energy transformation and building a new power system.

## 1. Introduction

With the deepening of the "double control" of carbon emissions and the photovoltaic plan, people's willingness of production enterprises to improve the efficiency of terminal energy use has become increasingly strong. It is an important choice for enterprises to transform their energy use mode by utilizing their resources and building a park-level integrated microgrid of photovoltaic charging and storage [1]. In the face of the next step, flexible resources such as massive photovoltaic and energy storage will be fully accessed. Power grid companies urgently need to solve two major problems, one is how to scientifically guide enterprises to carry out scientific and accurate photovoltaic and energy storage construction, to maximize the layout efficiency and utilization efficiency of enterprise terminal resources; the other is how to implement effective regulation and control strategies for enterprise terminal microgrid to meet the requirements of the government, power grid, and enterprises. The development of multi-energy complementary development and source-network-load-storage integration using distributed network structure is the inherent need to adhere to the concept of large-scale systems in the reform of the power industry. The objective requirement to realize the healthy and efficient development of the power system, and the task of improving the level of renewable energy development and the proportion of non-fossil energy consumption is imminent. It is of great significance to promote energy conversion and build a new modern power system.

## 2. Robust planning method for energy storage of photovoltaic microgrid

### 2.1. Set flexible resource planning index for PV microgrid

Photovoltaic microgrid power and grid load flexibility resources mainly include higher core grid, SOP, and controlled load [2]. The flexible supply of node resources and the coordination and cooperation of transmission channels provided by the network structure determines the flexible supply and demand balance of the photovoltaic microgrid. Based on the flexibility of space, time, direction, and adaptability of flexible resources, the balance index of flexibility supply and demand and the flexibility index of the node branch is proposed.

The flexibility demand of photovoltaic microgrids will change with the fluctuation of net load at any time. The so-called net load refers to the difference between the total load on each node of the photovoltaic microgrid and the total output capacity of the photovoltaic generator. The balance of supply and demand for flexible resources means that the microgrid can adapt to the requirements of the payload by collecting flexible resources supplied to higher or lower levels according to the actual situation [3]. The Equation for the supply and demand balance index $K_{FH}$ of the microgrid flexibility is:

$$K_{FH} = (K_{FH}^{up} + K_{FH}^{down}) / S \tag{1}$$

Where $K_{FH}$ represents the ratio between supply capacity and demand during the adjustment time interval. $K_{FH}^{up}$ and $K_{FH}^{down}$ represent the flexible supply-demand ratio of net load increase and decrease in a regulation interval. The supply and demand balance index of the flexibility resource in the microgrid is calculated by the Formula so that the regulation capability of the flexibility of the microgrid can be better improved according to the change of net load in the microgrid.

The supply of flexible resources between nodes must be provided by the grid system. As the main supply and guarantee channel of flexibility resources, the branch also needs to maintain a certain sense of load margin and can be adjusted in time according to the different net loads. In addition, because the allocation of flexibility resources in the microgrid is interfered with by different situations and generates randomness, increasing the branch load margin can also increase the utilization rate of flexibility resources in the microgrid and the ability to deal with unstable emergencies. A comprehensive analysis of the balance of branch load margin is helpful to better improve the regulation level of branches in the microgrid in response to sudden changes in net load [4]. Therefore, the calculation Formula of the node branch flexibility index $K_{CV}$ is:

$$K_{CV} = K_{CV1} + K_{CV2} \tag{2}$$

In this Equation, there is a negative correlation between $K_{CV}$ and load margin. The smaller the absolute value of $K_{CV}$ is, and the larger the absolute value of load margin is, the higher the load balancing degree of the branch is, and the safer the transmission path of flexibility resources is. $K_{CV1}$ reflects the value of the load ratio of the microgrid branch, and the load margin of the branch increases with the decrease of this value [5]. $K_{CV2}$ represents the capacity index of the load balance degree of the microgrid branch. The lower the value is, the more stable the load ratio is, and the stronger the capacity to deal with uncertain factors such as net charge. The capacity index of branch flexibility can be obtained through the above formulas so that the balance degree of the branch load margin and the branch load ratio is improved. The aim of flexibly dispatching the flexibility resources to the upper-level total network and the lower-level network structure by the flexibility resources in the microgrid when the net load is uncertainly changed is achieved.

*2.2. Calculate the best combination scheme of source-network load complementary*
The source-network load algorithm aims at planning the robust planning of photovoltaic microgrid energy storage and solves the coupling of distributed energy, energy storage, etc. With electric energy, it is to realize the planning of photovoltaic microgrid energy storage and improve the level of grid operation performance. According to the energy storage behaviors at the power supply side and node side of the microgrid, the energy storage characteristics are extracted and standardized. The correlation coefficient method is used to determine the degree of correlation between distributed flexibility resources and energy storage behaviors of photovoltaic microgrid [6]. For the negative correlation characteristics, the energy storage behavior is vectorized and the sum of the vectors is obtained. Then, the vector coordinates are simplified into a data set, and the optimal source-grid-load complementary scheme is obtained under the condition of the minimum standard deviation. The optimal planning scheme of the photovoltaic microgrid energy storage is obtained. In data processing, wield needs to standardize the data with different units and dimensions to ensure the authenticity, validity, and comparability of the data. The common processing methods include maximum-minimum standardization and decimal standardization. To facilitate the data after standardization to truly reflect the fluctuation trend of the data, the following methods are adjusted for this standardization:

$$z^* = \frac{z}{z_{\max}}$$

(3)

Where $z^*$ is the normalized value of the sample taken, $z$ is the original value of the sample, and $z_{\max}$ is the maximum of the sample. The load curve of the source-network load resources is selected for standardization, and the optimal combination scheme of multi-source complementary is calculated to maximize the complementary effect [7]. The purpose of solving the optimal complementary scheme is to minimize the standard deviation of the combined load data when different types of resources are combined, so as to make the curve the most stable. The result of the optimal complementary scheme is the optimal combination proportion of the capacity of different types of resource load. Therefore, it is necessary to determine the load behavior of the combined resources in the solution process.

*2.3. Build a photovoltaic microgrid energy storage robust configuration model*
Microgrid generally refers to a relatively independent power generation unit composed of some loads and some power generation units, which can generate electricity for its use. The microgrid has a variety of energy storage systems, and the model energy storage devices in this paper are mainly batteries and supercapacitors [8]. The general equivalent model of the battery is shown in Figure 1.

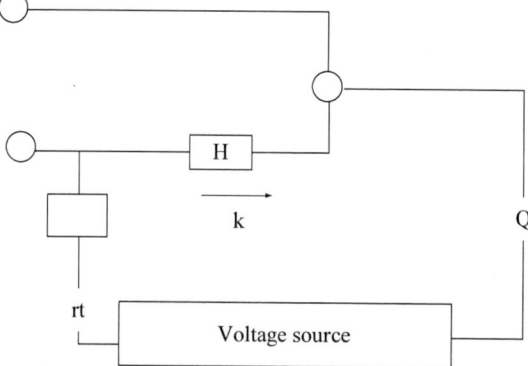

Figure 1. General equivalent model of battery

Its model is relatively simple, only consisting of a voltage source in series with a resistor. Like the battery, the state of charge (SOC) of the super-capacitor is also an important quantity to be recorded in the simulation model. The calculation Formula of the super-capacitor is as follows:

$$SOC_{mx,d} = (E_{mx,d} / E_{mx,g}) \times 100\%$$

(4)

Where $Emx, d$ represents the current capacity of the supercapacitor; $Emx, g$ refers to the rated capacity value of the capacitor. According to this formula, the capacity of the supercapacitor in the photovoltaic microgrid model is calculated to simulate the energy storage planning process [9]. At the same time, when constructing the model, after considering the robustness, the scenarios dealt with in the model will be worse. The demand for flexibility will increase. The capacity of the photovoltaic microgrid will be reduced by configuring energy storage for flexibility indicators, and the investment cost of energy storage will also increase.

The photovoltaic microgrid energy storage robust configuration model provided in this paper runs under the energy island model, so the microgrid needs to have energy that can maintain its own current and frequency stability at the same time. In this case, the microgrid energy storage can be checked. When the maximum load capacity of the power source distributed by the control model is reached, the terminal output voltage and frequency are constant regardless of the maximum load capacity of the model [10]. A dual closed-loop control strategy based on filter capacitor inner loop and flow feedback is indicated. The schematic diagram of the control system is shown in Figure 2.

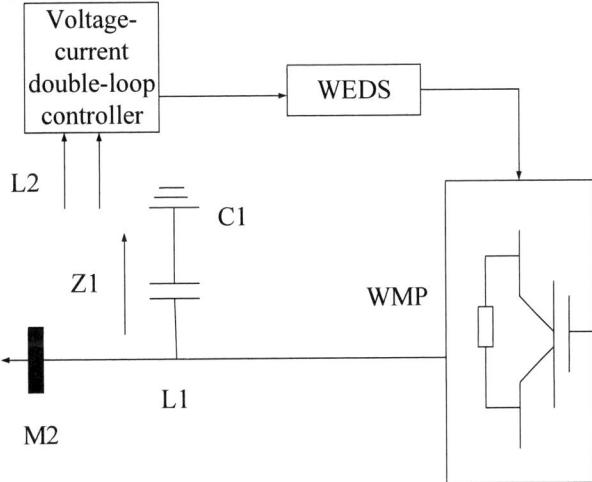

Figure 2. Schematic diagram of voltage inverter control

By adding a voltage inverter to the model, the model can maintain its own current and frequency stability energy during operation, and realize the increase of energy storage capacity more stably.

## 3. Experimental Test and Analysis

### 3.1. Preparation for the experiment

To prove the effectiveness of the proposed PV microgrid energy storage robust planning method, the PV microgrid energy storage system of a commercial building in a certain place is selected as the test object for the experiment.

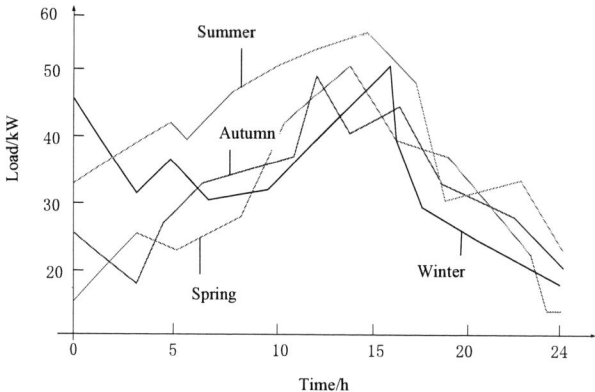

Figure 3. Typical load curve

As shown in Figure 3, the maximum load time in the curve of the maximum daily load in the four seasons is selected as the experimental time interval. The planning method proposed in this paper was used as the test object of the experimental group. Other two groups of traditional planning methods were introduced as control group 1 and control group 2 respectively to compare the changes of the three groups of methods in the energy storage process. In the course of the experiment, one hour is taken as an experimental time interval. 20 energy storage activities are carried out by three groups of methods, and the average value of each energy storage capacity is obtained by calculating the weighted average value. After the experiment, the average capacity of photovoltaic microgrid energy storage by three groups of methods is compared.

### 3.2. Experimental results

After 20 energy storage experiments, the experimental results are presented in Table 1.

Table 1. Experimental results of energy storage capacity of the PV microgrid

| Test groups | Total PV capacity/kW | The average value of storage capacity/Kw.h | The proportion of energy storage/% |
|---|---|---|---|
| Control group 1 | 206 | 67 | 32.52 |
| Control group 2 | 206 | 85 | 41.26 |
| Experimental group | 206 | 104 | 50.49 |

After the comparative experiment, the proportion of energy storage capacity of the experimental group reaches 50. 49% in the same energy storage time, while that of the control group 1 and 2 is 32. 52% and 41. 26%, respectively, which shows that the proposed energy storage planning method of photovoltaic microgrid is superior to the traditional energy storage planning method. The experiment is successful.

## 4. Conclusion

In this paper, a robust planning model for optimal allocation of energy storage in microgrid is constructed by setting the planning objective of flexible resource allocation in microgrid and calculating the optimal resource complementary scheme. Through the experiment, the method in this paper increases the proportion of energy storage capacity to 50. 49%, which is better than the traditional method. However, in the design process of this method, the analysis of source-network-load flexibility resources is not comprehensive enough. It is hoped that in the next study, the characteristics of source-network-load flexibility resources can be considered more comprehensively, and their role in the allocation of energy storage resources in photovoltaic microgrid can be maximized.

## References

[1] Zhu, X., Lu, G., Xie, W. (2021) Robust energy storage planning for distribution network considering source-load flexibility resources. Electric Power Automation Equipment, 41 (08): 8-16 + 40.

[2] Yu, Z., Liu, F., Liu, R. (2022) Review and prospect of source-load flexible resource optimization for distribution network elasticity improvement. China Electric Power, 55 (04): 132-144.

[3] Liu, M., Zheng, H., Qin, L. (2022) Research on integrated peak-shaving resource cooperation scheme based on source-network load storage. Electric Measurement and Instrumentation, 59 (08): 127-132.

[4] Zhang, N., Lu, J., Dai, H. (2020) Prospect of flexibility resources of China's power system under the coordinated development of source-network-load-storage. China Electric Power Enterprise Management, (16): 44-47.

[5] Liao, J., Wu, K., Liu, P. (2022) Power balance method for coordination of load and storage in new power system. Electrotechnical, (10): 132-138.

[6] Wang, H., Wang, Y., Chen, J. (2021) Optimal configuration of hybrid energy storage capacity in photovoltaic microgrid based on IPMOCSA. Journal of Electrical Engineering, 16 (03): 152-160.

[7] Wang, Z., Zhong, J., Zhang, Q. (2020) Energy management of photovoltaic microgrid based on hybrid energy storage under island operation. Electrotechnical Materials, (05): 48-51 + 55.

[8] Chen, J. (2020) Research on Capacity Optimal Allocation and Coordinated Control Strategy of Photovoltaic Storage DC microgrid. Shaanxi University of Science and Technology.

[9] Wang, Q., Sun, L., Li, J. (2021) Research on energy storage optimal configuration of building photovoltaic system based on phase change energy storage. Huadian Technology, 43 (09): 54-61.

[10] GAO, J., CHENm Z., ZHENG, X. (2021) Coordinated robust optimal allocation of multi-type energy storage in CCHP microgrid with photovoltaic. Journal of Electric Power Science and Technology, 36 (06): 56-66.

# Design and Research of the Intelligent Water Temperature Monitoring System

**Zhuolin Wang***

Department of Mechanical Engineering, School of Mechanical Engineering, Tsinghua University, Beijing, 100084, China

* Corresponding author: WangzhuolinTS@163.com

**Abstract.** In the industrial production process, people need to detect and control the temperature in various types of heating furnaces. The equipment selected for the water temperature control system designed in this paper is a single-chip microcomputer. The single-chip microcomputer has the characteristics of low power consumption, high performance, good reliability, and ease of production. Moreover, the single-chip microcomputer is more convenient, simple, and flexible in step control of temperature, which can improve the technical indicators of the controlled temperature, thus greatly improving the reliability of the product. In the design process, firstly, the hardware design is carried out. Secondly, the software design is carried out. And finally, comprehensive debugging is carried out to realize the constant temperature intelligent control of water temperature.

## 1. Introduction

At present, most water temperature control systems use transmission systems composed of analog temperature sensors, multi-channel analog switches, etc. This kind of system requires many temperature measurement cables to send the signals from the field sensors to the acquisition card, which is complex to install and disassemble and high in cost. At the same time, analog signals are transmitted on the line, which is vulnerable to interference and loss. And the measurement error is relatively large, which is not conducive to the controller making timely decisions based on temperature changes. In such a situation, it is necessary to develop a system with high real-time, high precision, and comprehensive processing of multi-point temperature information. The basic idea of this design is: to set a certain range of water temperature and automatically adjust when the ambient temperature changes to keep the set temperature basically unchanged.[1]

The system uses an 80C51 single-chip microcomputer as the controller. The forward channel is used for temperature acquisition and D/A conversion, and the backward channel is used for temperature control. The display channel is composed of LED. The working process of the system is as follows: firstly, the temperature sensor converts the temperature change into the corresponding electrical signal change, which means the temperature is converted into voltage and amplified, and then A/D conversion is carried out. This conversion converts the analog voltage into a binary digital voltage signal, which is transmitted to the 80C51 chip. The comparison and judgment with the set temperature range are realized through the program, and the temperature control is carried out according to the comparison results to maintain a constant water temperature. At the same time, the measured temperature is displayed by a digital tube. The executive part of the control circuit designed in this paper uses a LED for analog display. The system is set at 40 °C~90 °C (can be set according to actual needs). When the temperature is lower than 40 °C or higher than 90 °C, the LED lights up, indicating that the control circuit starts to work.[2]

## 2. Overall design scheme

The system adopts a combination of software and hardware. The basic idea is to program the required temperature range, convert the temperature into a digital signal using the hardware circuit, and transmit it to the microcontroller. The microcontroller compares the measured temperature with the set temperature and transmits the comparison results to the control circuit. The control circuit determines whether to work according to the received signal to maintain a constant temperature. Because the temperature range is written into the microcontroller, and the software determines whether the control circuit works or not, the error can be greatly reduced to a certain extent, and the operation is more convenient.[3]

The system is a typical detection and control application system, which can complete the whole process from water temperature detection, signal processing, and input calculation to output control and display. Therefore, a special computer application system should be formed with a single-chip microcomputer as the core to meet the functional requirements of control. In addition, the use of a single-chip microcomputer also provides the possibility of realizing the intelligent control of water temperature, providing a perfect man-machine interface and a multi-computer communication interface, which are difficult to achieve in conventional digital logic circuits.[4]

The main technical indicators of the system are as follows. The temperature setting range is 40°C $\sim$ 90 °C, and the minimum division is 1°C. The static error of temperature control: less than or equal to 1 °C. Two common anodes LED digital tube display, display temperature range: 35 °C~99 °C.

According to the system functions and design requirements, in order to simplify the system hardware, reduce the hardware cost, and improve the flexibility and reliability of the system, temperature calculation, nixie tube display, and most control processes can be completed by the software. The main functions of the hardware are temperature detection, output signal control, and temperature display. The block diagram of the overall system design scheme is shown in Figure 1.

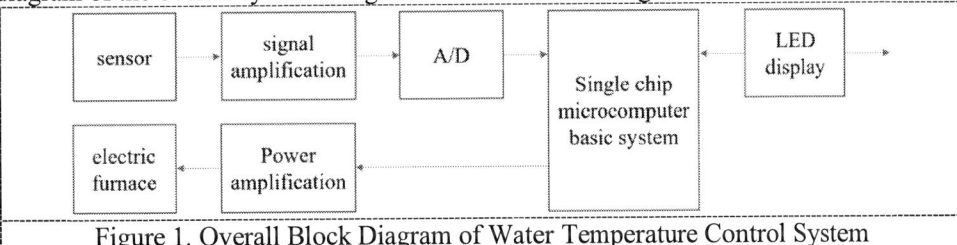

Figure 1. Overall Block Diagram of Water Temperature Control System

## 3. Unit circuit design

### 3.1. Forward passage

The forward channel is the channel for information acquisition, mainly including sensor detection, signal amplification, A/D conversion, and other circuits. Because the water temperature change is a relatively slow process, no sample and holds circuit is used in the forward channel. According to the design requirements, the static error of water temperature control is less than or equal to 1 °C. The water temperature setting range is 40 °C~90 °C, and the detection range of water temperature should be appropriately greater than this range, set as 35 °C~99 °C. Then, the total error of system control should not be greater than 1/(99-35) × 100%=1.56%, and the total error of signal acquisition allocated to the forward channel should not be more than half of the total system error, that is the accuracy should be 0.78%, which can be realized by 8-bit A/D converter, as shown in Figure 2.

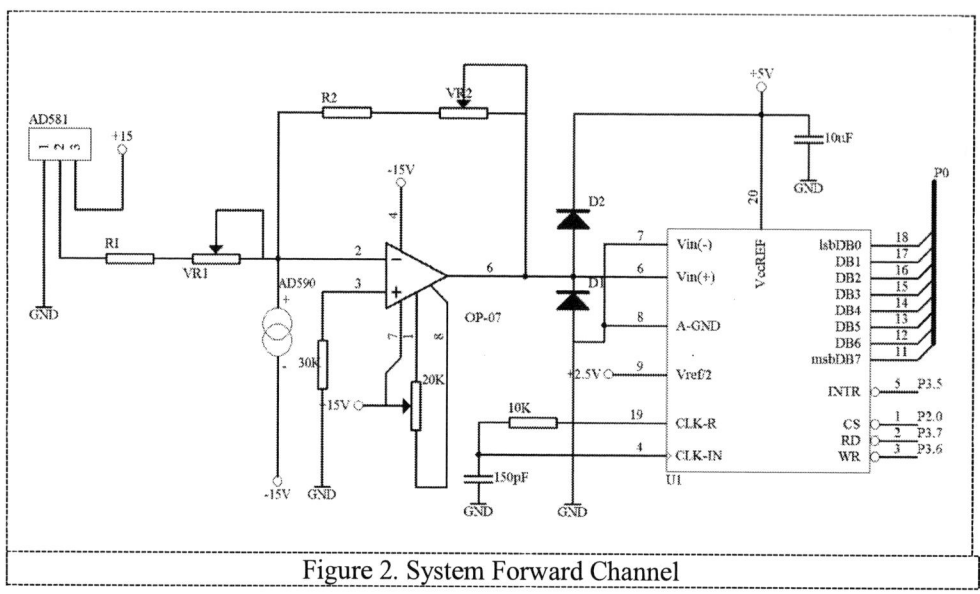

Figure 2. System Forward Channel

In Figure 2, the water temperature is sent to the input terminal of ADC0804 through the analog voltage signal of 0V~5V generated by the temperature sensor AD590 and the signal amplifier OP-07. ADC0804 converts the analog quantity into digital quantity and sends it to the microcontroller through the system bus for operation processing. The design of the forward channel includes the following aspects:

*3.1.1. Sensor Selection.* There are many kinds of temperature sensors. The thermocouple has low sensitivity due to its small thermoelectric potential; The precision of the thermistor is low due to its nonlinearity; the Platinum resistance temperature sensor is seldom used in small systems due to its high cost. AD590 is a two-terminal integrated temperature/current sensor produced by Analog Devices of the United States. It has a series of advantages such as small size, lightweight, good linearity, stable performance, etc. Its temperature measurement range is - 50 °C~+150 °C, and the error of the full-scale range is ± 0.3 °C. When the power supply voltage is between 5V~10V and the stability is 1%, the error is only ± 0.01 °C, which is completely suitable for the requirements of this design for water temperature measurement. In addition, AD590 is a temperature/current sensor, which is also helpful in improving the anti-interference ability of the system. Therefore, AD590 is selected as the temperature sensor in this design.[5]

It should be noted that when using sensors such as AD590, in order to avoid direct contact between the device and the measured liquid, the sensor should be installed into a protective sleeve, or the device should be sealed with materials such as polytetrafluoroethylene, hard vinyl resin, etc., to avoid the corrosion of the measured liquid on the sensor and the impact on the measurement accuracy.

*3.1.2. Signal conversion and amplification circuit.* In Figure 2, the three terminal voltage regulator AD581 provides 10V standard voltage. It, together with the operational amplifier and resistors R1, VR1, R2, and VR2, forms a signal conversion and amplification circuit, which converts the temperature of 35 °C to 99 °C into a voltage signal of 0V to 5V and amplifies it. Because the water temperature changes relatively slowly, the signal conversion and amplification circuit have no requirements for the bandwidth of the operational amplifier. On the other hand, the output current of the AD590 is 308.2uA and 372.2uA at 35 °C and 99 °C, respectively, while the input offset current and zero drift of the operational amplifier are relatively small, which can be ignored. Therefore, a general-purpose operational amplifier OP-07 can be used.[6]

### 3.1.3. A/D converter

Analog-to-digital converter (ADC for short) is used to convert analog quantity into digital quantity. The n-bit analog-to-digital converter outputs an n-bit binary number, which is proportional to the analog voltage applied to the input. There are many methods to realize analog-to-digital conversions, such as parallel ADC, successive approximation ADC, and double integral ADC. Parallel ADC is the fastest, but its cost is too high, and its precision is not high. Double integral ADC has high precision and strong anti-interference ability, but its speed is too slow, which is suitable for converting slowly changing signals. Successive approximation ADC has the advantages of high conversion accuracy, medium working speed, and low cost, so it has been widely used.[7]

In this design, because the total error of the forward channel is 0.78%, and the system also has low requirements for the speed of signal acquisition, the low-cost 8-bit successive approximation A/D converter ADC0804 is selected. The conversion speed of the converter is 100us, the conversion accuracy is 0.39%, and the corresponding error is 0.234 ℃.

The signal connection of ADC0804 is shown in Figure 2. Wherein a resistor and a capacitor are externally connected at both ends of CLK-R and CLK-IN to generate the clock signal required for A/D conversion. Chip selection is controlled by P2.0 of 8051. The INTR of the A/D converter is connected to P3.5 of 80C51, and the microcontroller obtains the information about the completion of the A/D converter conversion by querying.

### 3.2. Single-chip microcomputer basic system

As shown in Figure 3, the basic system of the single-chip microcomputer is the core of the whole control system, which completes the information processing and coordinated control functions of the whole system. Compare and judge the conversion value of the read-in temperature with the set temperature value, output different control signals according to the results, and convert the measured temperature value into decimal numbers for display. Because the system has no special requirements for control speed, accuracy, and function, the MCS-51 series single-chip microcomputer 80C51, which is widely used at present, can be selected.

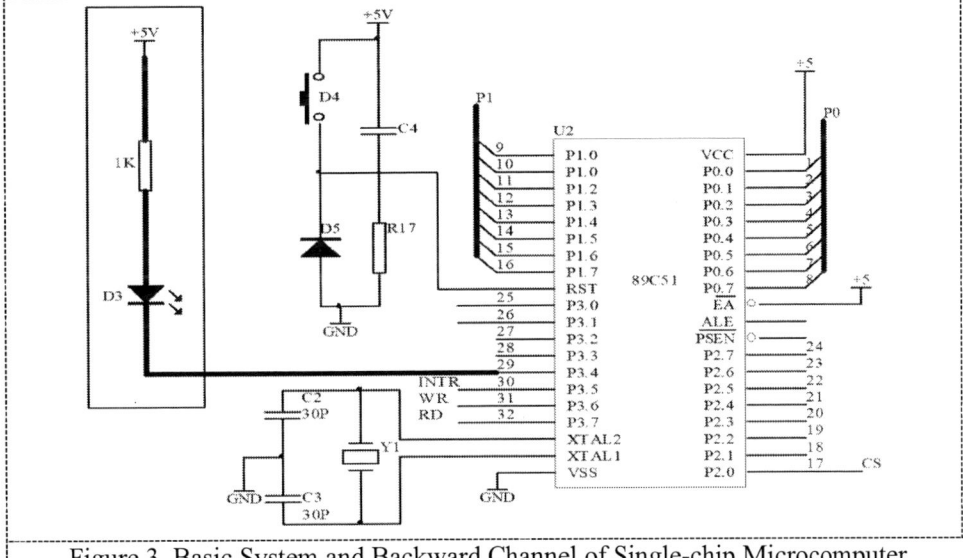

Figure 3. Basic System and Backward Channel of Single-chip Microcomputer

This design takes the single-chip microcomputer as the basic system and the MCS-51 series single-chip microcomputer 80C51 as the core. 80C51 is an 8-bit (8-bit data line) single-chip microcomputer with 256BRAM and 4KBEPROM on the chip. The central processor unit realizes the operation and control functions. There are 256 units in the internal data memory, and the addresses to access them are 00H~FFH. The user uses the first 128 units (00H~7FH), and special function registers occupy the last

128 units. The internal 2 16-bit timing/counters are used for timing or counting. The control function can be realized by timing or counting results. 80C51 has four 8-bit parallel ports (P0, P1, P2, P3) for address output and data input/output. There is also a clock oscillator in the chip, which can be oscillated only by connecting quartz crystal outside. 80C51 adopts 40 pins dual in-line package (DIP).

### 3.3. Backward passage

The backward channel, as shown in the wireframe at the upper left corner of Figure 3, is the channel to realize control signal output. According to the total system error requirements, the control accuracy of the backward channel should also be controlled within 0.78%. In this design, the backward channel is simulated and displayed by a LED. When the temperature is lower than or higher than the measured range, the LED will emit light; When the temperature is within the measured range, the diode goes out. For this design, when the measured temperature is between 40 °C and 90 °C, the LED is dark, and when the measured temperature is greater than 90 °C or less than 40 °C, the LED is bright.[8]

### 3.4. Display channel

As shown in Figure 4, the display channel is mainly composed of an LED display composed of two digital tubes, which display the measured temperature with a display range of 35 °C~99 °C. LED digital tube, also known as the digital semiconductor tube, is the most used display device in the current digital circuit. It is made of light-emitting diodes as pen segments and sealed in the form of a common cathode or a common anode. In this design, P3.0 controls one bit, P3.1 controls ten bits, and the nixie tube uses a common anode.

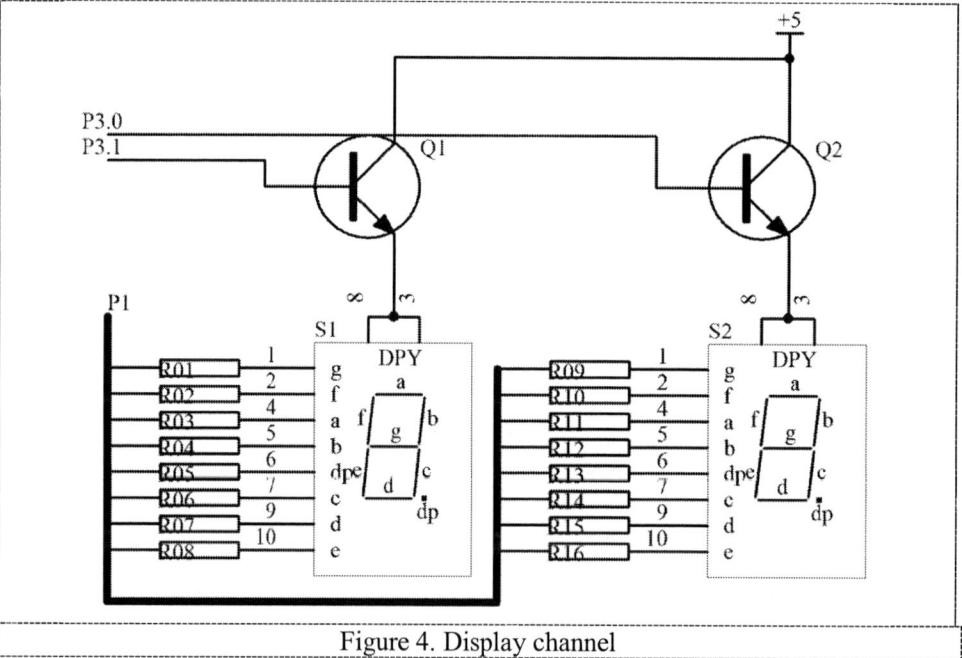

Figure 4. Display channel

## 4. Software design

With 80C51 as the core, the P0 port is the signal input port, the P1 port is the signal output port, and P3.4 is the output control port. First, read the signal output from ADC0804, convert the input signal to the corresponding decimal value by using the appropriate calculation method, then display the measured temperature on the nixie tube, and then compare the measured temperature with the set temperature to determine the output of the corresponding control signal. The procedure flow is shown in Figure 5.

The conversion module converts the digital signal provided by ADC0804 into a decimal value, and the display module displays the converted decimal value on the nixie tube. The nixie tube selects a common anode and uses a dynamic display, displaying one digit first and then ten digits. P3.0 controls the display of one digit of the nixie tube, and P3.1 controls the display of ten digits. When P3.0 is high level, P3.1 is low level, and P3.1 is high level, P3.0 is low level, and P3.0 is low level, it gates the nixie tube representing ten digits.

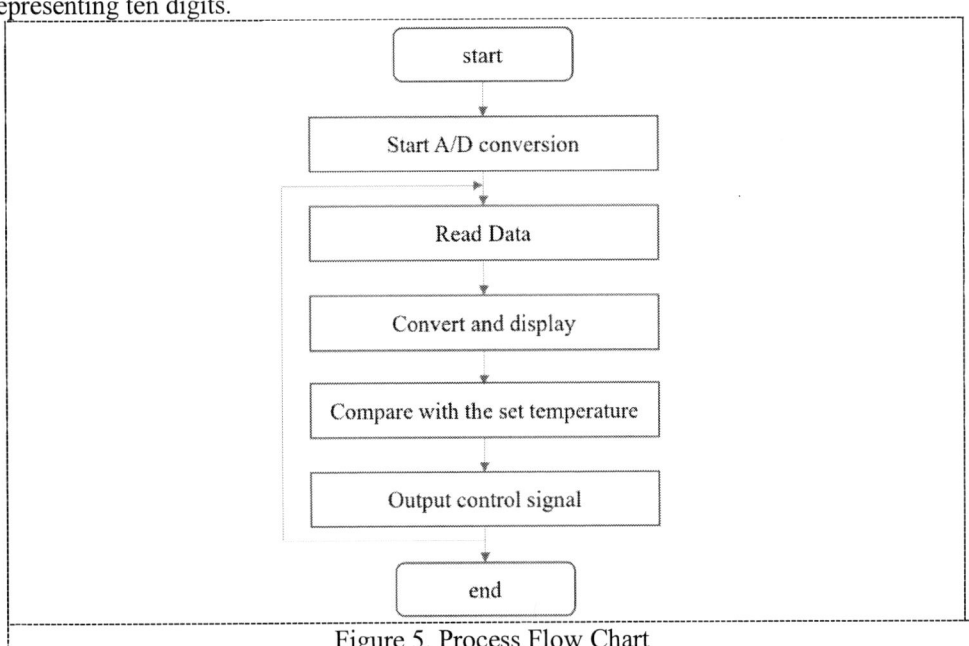

Figure 5. Process Flow Chart

The comparison output module compares the converted decimal value with the set temperature range of 40 °C~90 °C. If it is between 40 °C and 90 °C, P3.4 outputs a high level, and the LED is dark; If it is greater than 90 °C or less than 40 °C, P3.4 outputs a low level, and the LED lights up.

## 5. Hardware debugging

The basic system of MCU, display channel, forward channel, and backward channel, are debugged, respectively. During debugging, the simulator can be used to read and write the interface address and statically test whether the connection of each part of the circuit is correct; For the dynamic process, we can write a short debugging program to coordinate with the debugging of the hardware circuit.

### 5.1. Basic system debugging of the single-chip microcomputer

*5.1.1. Crystal oscillator circuit.* Turn the crystal oscillator switch of the emulator to the outside. If the emulator crashes, it indicates that there is a problem with the crystal oscillator of the user system. Currently, use an oscilloscope to observe whether there is an oscillation signal at the clock signal input end of the microcontroller, or check the parameters of each component of the crystal oscillator circuit.[9]

*5.1.2. Reset circuit.* Press the reset button to put the system in the reset state, or use a multimeter to check the signal and device parameters at each point of the reset circuit.

### 5.2. Commissioning of forward channel

*5.2.1. Static working point debugging.* Heat the water temperature and test it with a thermometer. When the water temperature is 35 °C, adjust the VR1 resistance to make the output voltage of the amplifier OP-07 0V. When the water temperature is 99 °C, adjust the VR2 resistance value to make OP-07 output 5V. Test the output voltage of the op-amp OP-07 at any number of points within the range of 35 °C~99 °C.

*5.2.2. A/D converter commissioning.* Select several test points within the range of 35 °C~99 °C, and use the simulator to write any number to ADC0804 to start A/D conversion. Read the conversion result from ADC0804 and compare it with the test value. The result is incorrect. Check whether the connection between ADC0804 and 80C51 is correct and whether the reference voltage of ADC0804 is+5V.

*5.3. Backward channel commissioning*

*5.3.1. Static debugging.* Use the simulator to output a low level on P3.4, and the LED will turn on, and output a high level on P3.4, and the LED will turn off. If the output is abnormal, check the connection and welding of each part according to the signal output sequence.

*5.3.2. Dynamic debugging.* In the system design, P3.4 controls the output. When the temperature is higher than 90 °C or lower than 40 °C, P3.4 should output a low level. Write a short debugging program, periodically output a certain duty cycle PWM waveform on P3.4, and observe the light and dark conditions of the diode with an oscilloscope.

## 6. Translator Simulation

ZHUANH:
CLR C
MOV B, #4 ;   4 Put in register B
MOV A, R0 ;   R0 is put into register A
ADD A, #1
JNC ZHUANH1 ;   Jump to ZHUANH1 if carry is not 1
MOV R0, #99
AJMP ZHUANH2 ;   Jump directly to ZHUANH2
ZHUANH1:
DIV AB
ADD A, #35
MOV R0, A
RET
ZHUANH2: RET

Set it to 00000000 H (decimal 0) at 35 °C, 11111111H (decimal 240) at 99 °C. If the difference between 35 °C and 99 °C is 64 °C, the decimal value is corresponding to the measured digital quantity is D, and the temperature is T, then T=(D+1)/4+35. Simulation process: if R0 is set to 00010011, it can be seen from the Table that it is decimal 19, T=(19+1)/4+35=40, and the temperature displayed through calculation should be 40 °C. The 51 series simulation system is used to directly set 00010100 to R0 on the computer, and the calculation result is 40 °C, indicating that the simulation program is correct.[10]

## 7. Conclusion

This design is a single-chip microcomputer control system. The temperature can be kept within a certain range, and the system is stable and reliable. It has great value in daily life, industrial and agricultural production, and scientific experiments. Single-chip microcomputer technology makes traditional temperature control intelligent. Because the function of the temperature control system is controlled by software, the temperature control algorithm can be easily adjusted according to the application situation to meet the requirements. In addition, with the progress of microelectronics technology, large-scale ASIC is preferentially selected in the design of temperature control systems, which can make the hardware clear and simple, compress the device volume, greatly reduce the error, effectively improve the accuracy, and anti-interference performance of the device.

## References

[1] Ma Y, Ding W. (2018) Design of intelligent monitoring system for aquaculture water dissolved oxygen. IEEE 3rd Advanced Information Technology, Electronic and Automation Control Conference (IAEAC). IEEE, 414-418.

[2] Lakhiar I A, Jianmin G, Syed T N, et al. (2018) Monitoring and control systems in agriculture using intelligent sensor techniques: A review of the aeroponic system. Journal of Sensors.

[3] Huan J, Li H, Wu F, et al. (2020) Design of water quality monitoring system for aquaculture ponds based on NB-IoT. Aquacultural Engineering, 90: 102088.

[4] Niswar M, Wainalang S, Ilham A A, et al. (2018) IoT-based water quality monitoring system for soft-shell crab farming. IEEE International Conference on Internet of Things and Intelligence System (IOTAIS). IEEE, 6-9.

[5] Komarenko O, Hrytsyk I. (2019) PSOC4 BASED INTELLIGENT WATER CONSUMPTION METER.

[6] Saravanan K, Anusuya E, Kumar R, et al. (2018) Real-time water quality monitoring using Internet of Things in SCADA[J]. Environmental monitoring and assessment, 190(9): 1-16.

[7] Shi B, Sreeram V, Zhao D, et al. (2018) A wireless sensor network-based monitoring system for freshwater fishpond aquaculture[J]. Biosystems engineering, 172: 57-66.

[8] Stewart R A, Nguyen K, Beal C, et al. (2018) Integrated intelligent water-energy metering systems and informatics: Visioning a digital multi-utility service provider[J]. Environmental Modelling & Software, 105: 94-117.

[9] AshifuddinMondal M, Rehena Z. (2018) Iot based intelligent agriculture field monitoring system. 8th International Conference on Cloud Computing, Data Science & Engineering (Confluence). IEEE, 625-629.

[10] Koditala N K, Pandey P S. (2018) Water quality monitoring system using IoT and machine learning. International Conference on Research in Intelligent and Computing in Engineering (RICE). IEEE, 1-5.

# Harmonic Voltage Amplitude Calculation and Fundamental Amplitude Correction Based on Synchronized SVPWM of Two-level VSI under Low Switching Frequency

**Yueqiang Hu[1,a*], Fengguang Jiang[2,b]**

[1]School of Artificial Intelligence and Automation, Huazhong University of Science and Technology, Wu Han, Hu Bei, China

[2]Wuhan Zhengyuan Electric Co, Wu Han, Hu Bei, China

[a]email: m202173139@hust.edu.cn, [b]email: fengguang1988@qq.com

**Abstract.** The switching frequency of two-level VSI is usually limited to a very low level for high-power drives to minimize the switching losses. For the synchronized SVPWM under low switching frequency, there will be a relative deviation of fundamental amplitude between the output voltage and the reference voltage. To solve this problem, in this paper, a method to calculate the amplitude of output phase voltage harmonics under synchronized SVPWM is proposed, and an online fundamental amplitude correction method is proposed based on this calculation result.

## 1. Introduction

In some high-power, high-frequency applications, such as locomotive traction systems, the switching frequency of two-level VSI must be restricted to a low level to reduce the switching losses of the switching devices. Under low switching frequencies, synchronized modulation is often used to ensure the synchronization and symmetry of the output waveform [1]. The representative synchronized modulation methods are selected harmonic elimination PWM (SHEPWM) and synchronized SVPWM. SHEPWM incorporates fundamental wave amplitude into the computational equation of the optimization criterion and offers desired control of fundamental voltage under low switching frequency [2]. However, it faces the transcendental equation, which makes SHEPWM complicated [3]. Synchronized SVPWM follows the volt-second balance principle, and this method calculates the proportion of each vector online. By reasonably arranging the switching sequences and modulation positions, synchronous and symmetric modulation is made by synchronized SVPWM. But in the case of low switching frequency, the fundamental amplitude of output voltage has a relative deviation from the reference voltage. This paper gives a method to calculate phase voltage harmonics by using synchronized SVPWM under low switching frequency and proposes an online fundamental voltage amplitude correction method based on this calculation result.

## 2. Principles and Methods

The topology of the two-level VSI is shown in Fig.1. Defining a set of bridge arms with the upper switch tube on and the lower switch tube off as 1 and the lower switch tube on and the upper switch tube off as 0. There are 8 vectors, corresponding to 6 basic vectors and two zero vectors, which are labeled as $V_0(000)$, $V_1(100)$, $V_2(110)$, $V_3(010)$, $V_4(011)$, $V_5(001)$, $V_6(101)$, $V_7(111)$.

Content from this work may be used under the terms of the Creative Commons Attribution 3.0 licence. Any further distribution of this work must maintain attribution to the author(s) and the title of the work, journal citation and DOI.
Published under licence by IOP Publishing Ltd

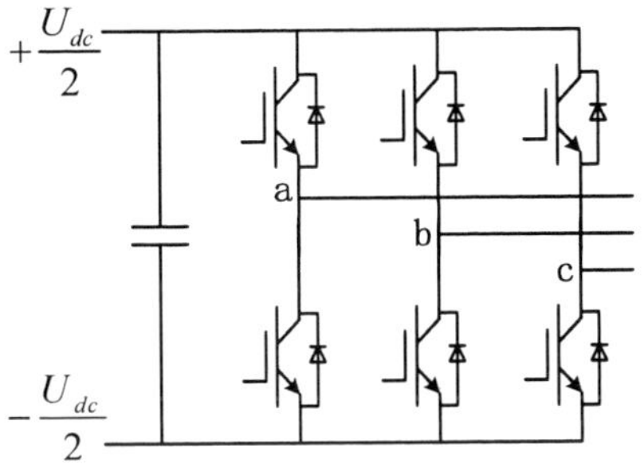

Fig.1 The topology of the two-level voltage inverter

Taking the conventional space vector strategy with a carrier ratio of 9 (CSVS9) for example, according to [4], synchronized SVPWM can be achieved by setting the sampling points and switching sequences in Table 1, which makes the output waveform be three-phase symmetric, 1/4-cycle even symmetric, and half-wave odd symmetric. The sampling points and switching sequences of CSVS9 are shown in Table 1, and for the sake of description easily, $V_0$, $V_1$...$V_7$ are abbreviated as 0, 1...7.

Table 1 The sampling points and switching sequences of CSVS9

| Position of samples | 10° | 30° | 50° | 70° | 90° | 110° | 130° | 150° | 170° |
|---|---|---|---|---|---|---|---|---|---|
| Vector Sequences | 7210 | 0127 | 7210 | 0327 | 7320 | 0327 | 7430 | 0347 | 7430 |
| Position of samples | 190° | 210° | 230° | 250° | 270° | 290° | 310° | 330° | 350° |
| Vector Sequences | 0547 | 7450 | 0547 | 7650 | 0567 | 7650 | 0167 | 7610 | 0167 |

Because of the symmetry, only one of three phases is needed to complete the calculation of the output voltage harmonics. Fig.2 shows the schematic diagram of a-phase modulation. Using the concept of switching angles in SHEPWM, the four switching angles under CSVS9 can be derived based on the switching action moments within 270°~360°. The steps are as follows.

### 2.1 Calculating the comparison value $T_{cma}$ of phase a

Suppose a modulated waveform generates $T_{cma}$ as shown in Fig.2. The average distribution of two zero vectors can effectively reduce the output harmonic content, so here, we divide the dwell time of two zero vectors equally.

### 2.1.1 $T_{cma\_270°}$

When reference vector $V_{ref}$ located at 270°, the vector $V_5$ and $V_6$ dwell for equal time, so it can be deduced that:

$$T_{cma\_270°} = T/2 \qquad (1)$$

Where $T = 1/(18 * f_s)$, $f_s$ is the frequency of $V_{ref}$.

### 2.1.2 $T_{cma\_290°}$

When reference vector $V_{ref}$ located at 290°, Fig.3 shows the vector synthesis process at 290°. According to the volt-second balance, the dwell time of $V_5$ and $V_6$ can be derived:

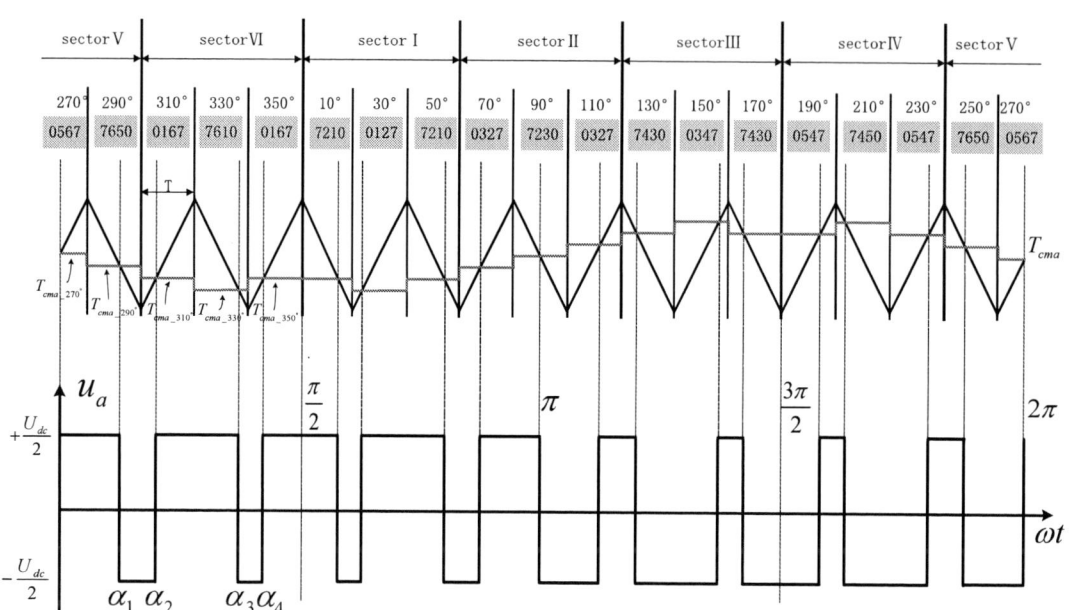

Fig.2 Schematic diagram of phase a modulation

$$V_{ref} \cdot T = V_5 \cdot T_1 + V_6 \cdot T_2 \tag{2}$$

Where $T_1$, $T_2$ are dwell time of $V_5$, $V_6$, respectively. By coordinating decomposition, it yields:

$$|V_{ref}| \cos(7\pi/18) \cdot T = U_{dc} \cdot T_2/3 - U_{dc} \cdot T_1/3 \tag{3}$$

$$|V_{ref}| \sin(7\pi/18) \cdot T = U_{dc} \cdot T_2/\sqrt{3} + U_{dc} \cdot T_1/\sqrt{3} \tag{4}$$

$T_1$, $T_2$ are calculated by:

$$T_1 = m \sin(7\pi/18) \cdot T/\sqrt{3} - \mathrm{m} \cdot \cos(7\pi/18) \cdot T \tag{5}$$

$$T_2 = m \sin(7\pi/18) \cdot T/\sqrt{3} + \mathrm{m} \cdot \cos(7\pi/18) \cdot T \tag{6}$$

Where $m$ is the Modulation ratio and $m = |V_{ref}|/(2U_{dc}/3)$, $T_{cma\_290°}$ is the sum of the dwell time of $V_0$ and dwell time of $V_5$:

$$T_{cma\_290°} = (T - T_1 - T_2)/2 + T_1 \tag{7}$$

### 2.1.3 $T_{cma\_310°}$

When reference vector $V_{ref}$ located at 310°, Fig.4 shows the vector synthesis process at 310°, which is the same as (2)

$$V_{ref} \cdot T = V_1 \cdot T_4 + V_6 \cdot T_3 \tag{8}$$

Where $T_3$ and $T_4$ are dwell time of $V_6$, $V_1$, respectively. $T_3$, $T_4$ are calculated by

$$T_3 = 2m \sin(5\pi/18) \cdot T/\sqrt{3} \tag{9}$$

$$T_4 = m \cos(5\pi/18) \cdot T - m\sin(5\pi/18) \cdot T/\sqrt{3} \tag{10}$$

$T_{cma\_310°}$ is equal to the dwell time of $V_0$

$$T_{cma\_310°} = (T - T_3 - T_4)/2 \tag{11}$$

### 2.1.4 $T_{cma\_330°}$ and $T_{cma\_350°}$

According to section 2.1.3, $T_5$, $T_6$, $T_7$ and $T_8$ are calculated by

$$T_5 = 2m \sin(3\pi/18) \cdot T/\sqrt{3} \tag{12}$$

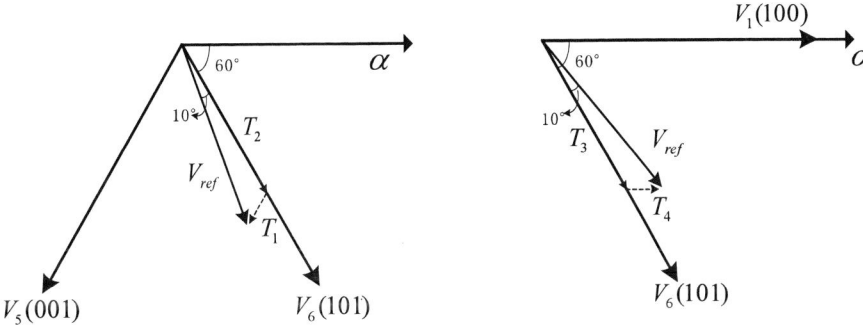

Fig.3 Vector synthesis process at $290°$    Fig.4 Vector synthesis process at $310°$

$$T_6 = m\cos(3\pi/18) \cdot T - m\sin(3\pi/18) \cdot T /\sqrt{3} \tag{13}$$

$$T_7 = 2m\sin(\pi/18) \cdot T /\sqrt{3} \tag{14}$$

$$T_8 = m\cos(\pi/18) \cdot T - m \cdot \sin(\pi/18) \cdot T /\sqrt{3} \tag{15}$$

Where $T_5$ and $T_6$ are dwell time of $\boldsymbol{V_6}, \boldsymbol{V_1}$, respectively, while reference vector $\boldsymbol{V_{ref}}$ located at $330°$. $T_7, T_8$ are dwell time of $\boldsymbol{V_6}, \boldsymbol{V_1}$, respectively, while reference vector $\boldsymbol{V_{ref}}$ located at $350°$.

$T_{cma\_330°}$ and $T_{cma\_350°}$ are equal to the dwell time of $\boldsymbol{V_0}$

$$T_{cma\_330°} = (T - T_5 - T_6)/2 \tag{16}$$

$$T_{cma\_350°} = (T - T_7 - T_8)/2 \tag{17}$$

According to the symmetry, we can obtain the remaining comparison values.

### 2.2 Calculating the switch angle α of phase a

According to Fig.2, four switching angles can be calculated as follows.

$$\alpha_1 = \left(1.5T - T_{cma\_290°}\right) * \pi/(9T) \tag{18}$$

$$\alpha_2 = \left(1.5T + T_{cma\_310°}\right) * \pi/(9T) \tag{19}$$

$$\alpha_3 = \left(3.5T - T_{cma\_330°}\right) * \pi/(9T) \tag{20}$$

$$\alpha_4 = \left(3.5T + T_{cma\_350°}\right) * \pi/(9T) \tag{21}$$

### 2.3 Calculating the fundamental and harmonic content

After calculating the switching angles, Fourier decomposition is carried out. In fact, due to the symmetrical design, we do not need to consider the calculation of even harmonics.

$$u = \sum_{n=1,3,5\ldots}^{\infty} \frac{2U_{dc}}{n\pi}\left[1 + 2\sum_{i=1}^{N}(-1)^i \cos(n\alpha_i)\right]\sin(n\omega t) \tag{22}$$

Where $n$ is the $n$th harmonic, $n = 1,3,5,7\ldots$, $U_n$ is the amplitude of $n$th harmonic, $U_{dc}$ is the dc-link voltage, $\omega$ is the fundamental frequency, $N$ is the number of switching angles in 1/4 cycle, and is equal to 4 here, and $\alpha_i$ is the $i$th switching angle.

### 2.4 Fundamental amplitude correction

Once the output fundamental amplitude $U_1$ is obtained, the output fundamental can be corrected. The method is as shown in Fig.5. Putting the difference value between the amplitude of the reference voltage $\boldsymbol{V_{ref}}$ and the fundamental amplitude $U_1$ calculated by section 2.3, which is called $|V_{ref\_e}|$ into PI module, the fundamental amplitude compensation signal $|V_{ref\_c}|$ is generated. The sum of $|V_{ref\_c}|$ and reference voltage amplitude $|V_{ref}|$ is the final modulating voltage amplitude $|V_{ref\_l}|$, which is used for synchronized SVPWM and generates gate driving signal $g$ to drive VSI.

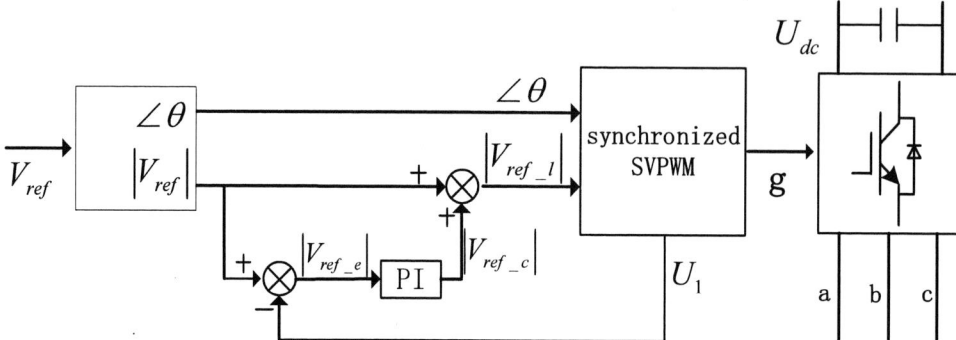

Fig.5 Fundamental amplitude correction method

## 3. Simulation Results

In this section, the proposed method is simulated in the environment of MATLAB/Simulink. The simulation parameters are: DC bus voltage $V_{dc}$= 750V, the amplitude of reference voltage is 350V, frequency = 50Hz, with CSVS9 synchronized SVPWM using.

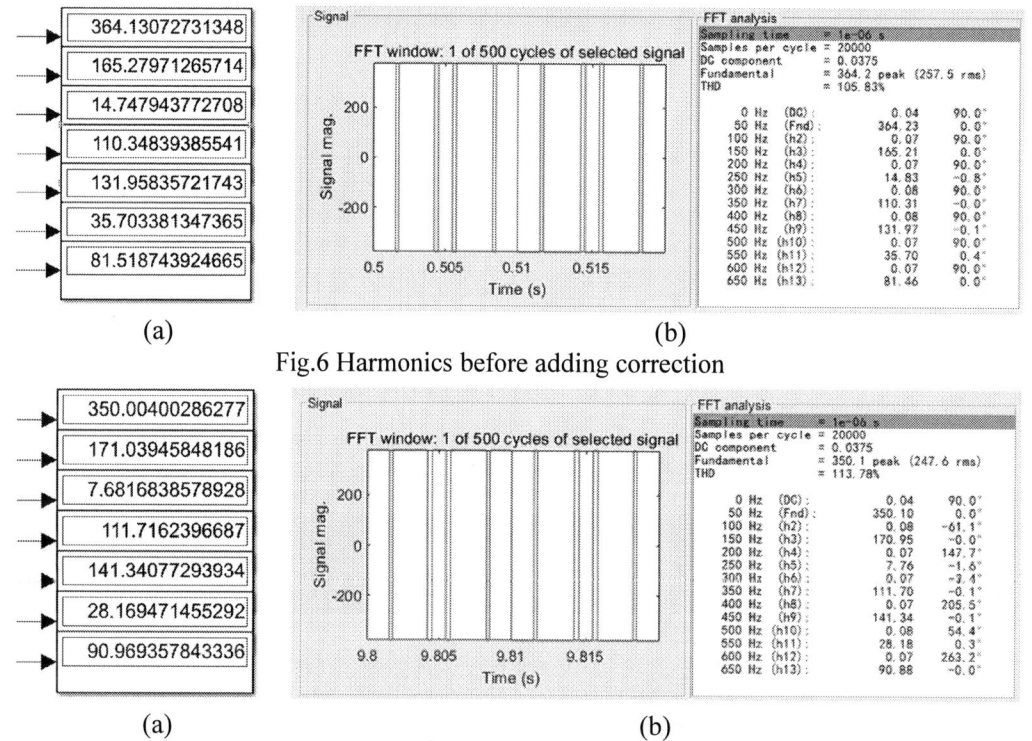

Fig.6 Harmonics before adding correction

Fig.7 Harmonics after adding correction

The calculated values of harmonics before adding correction are shown in Fig.6(a), and the results of Simulink's FFT analysis before adding correction are shown in Fig.6(b). Correction added at 1 s, The calculated values of harmonics after adding correction are shown in Fig.7(a), and the results of Simulink's own FFT analysis after adding correction are shown in Fig.7(b). Calculated values from top to bottom are fundamental, 3rd harmonic, 5th harmonic, ..., and 13th harmonic. As we can see that the harmonic content calculated by the proposed calculation method is almost equal to the results of the FFT analysis that comes with Simulink, the calculation is correct. From Fig.8, it can be seen that the error signal of the fundamental amplitude $|V_{ref\_e}|$ is gradually corrected to 0 after adding correction, correction is effective.

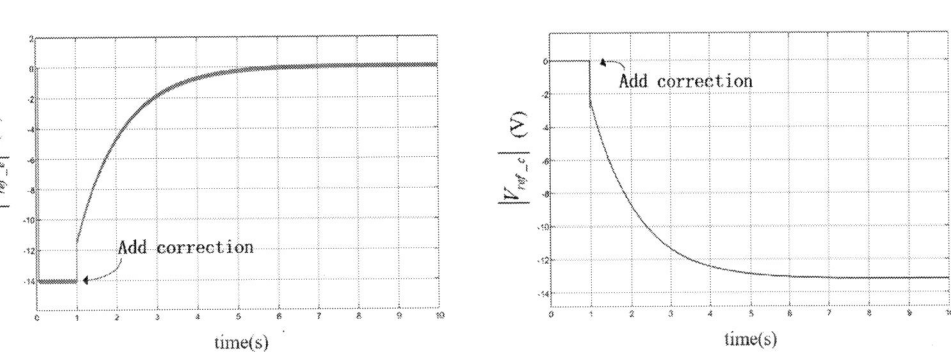

Fig.8 (a) $|V_{ref\_e}|$ Waveform variation  (b) $|V_{ref\_c}|$ Waveform variation

## 4. Conclusions

This paper focuses on the fundamental amplitude deviation caused by the synchronized SVPWM of two-level VSI under low switching frequency. Firstly, a method to calculate the content of each harmonic of the output voltage is proposed, from which the fundamental amplitude can be calculated. Then, a method to correct the fundamental amplitude deviation online is proposed. Simulation shows that the harmonic calculation method is accurate, and the fundamental amplitude correction method is simple and feasible.

## References

[1] Zhang, G., Zhou, Z., Shi, T., Xia, C. (2021) An Improved Multimode Synchronized Space Vector Modulation Strategy for High-Power Medium-Voltage Three-Level Inverter. IEEE Transactions on Power Electronics., vol. 36, no. 4, pp. 4686-4696.

[2] Rai, N., Chakravorty S. (2020) A Review on the Generalized Formulations for Selective Harmonic Elimination (SHE-PWM) strategy. In: IEEE First International Conference on Smart Technologies for Power, Energy and Control (STPEC). Nagpur. pp. 1-6.

[3] Yang, H., Zhang, Y., Yuan, G., Walker, P. D., Zhang, N. (2017) Hybrid Synchronized PWM Schemes for Closed-Loop Current Control of High-Power Motor Drives. IEEE Transactions on Industrial Electronics., vol. 64, no. 9, pp. 6920-6929.

[4] Wang, K., You, X., Wang, C., Zhou, M. (2015) Research on Synchronized SVPWM Strategies Under Low Switching Frequency. Proceedings of the CSEE.,35(16).4175-4183.

# On-line Monitoring of Instantaneous Over-voltage Intelligent Monitoring of EMUs

**Zhongjiang Sun[1,a], Suyue Liu[1, b*], Zhiming Yang[1,c] and Shuangfeng Xu[1,d]**

[1.] CRRC Qingdao Sifang Rolling Stock Research Institute Co., Ltd., Qingdao, China, 266000

[a]Email: zhongjiangsun@163.com, [b*]E-mail: sarah.suyue@foxmail.com, [c]E-mail: zhimingyang0728@163.com, [d]E-mail: shuangfeng_xu@foxmail.com

**Abstract:** Since high-speed EMUs have high requirements for the reliability of the traction power supply system and the safety of train operation, the online monitoring and analysis of transient over-voltage are of great significance. By analyzing and comparing the characteristics of the existing power grid over-voltage online monitoring devices at home and abroad, and due to the difficulty in detecting the power grid voltage on the high-speed train, a new type of over-voltage online monitoring using the broadband transmission characteristics of the equipment is proposed for the high-speed train. The principle of the scheme is to reconstruct the real-time voltage waveform of the primary side of the transformer by convolving the secondary side voltage waveform of the vehicle-mounted voltage transformer through the fast Fourier transform with the transformer-specific transmission parameter model and to correlate the relevant data, such as power, voltage fluctuation, harmonic distortion, etc. Simultaneously, simulation and experiments are carried out to verify the validity and stability of the scheme algorithm. The experimental results show that the system has good response characteristics, high measurement accuracy, and stable and reliable working conditions. The system achieves the expected design goals, and it has been successfully applied to the online monitoring equipment of EMU power quality. While monitoring the over-voltage of the power grid, this system provides data support for the research on the aging and life of the onboard high-voltage electrical equipment.

## 1. Introduction

In the modern power supply system, over-voltage often occurs. Over-voltage of the power grid may be caused by natural conditions such as lightning stroke, fault, resonance, or personnel operation.[1] However, once the electrical equipment in train operation encounters over-voltage, accidents such as breakdown, discharge, flashover, and explosion are likely to happen, which will lead to disorder of onboard sensitive equipment control system, aging of on-board high-voltage electrical equipment, thus giving rise to major potential safety hazards that accelerate the damage or explosion of on-board high-voltage electrical equipment.[2] Thereby, the monitoring and analysis of over-voltage in the EMU power system are of great significance. In addition to the over-voltage generated by the unstable grid voltage, the operating over-voltage caused by neutral-section passing and start-stop of EMUs also has a great impact on the onboard voltage transformer and the running train.[3]

On-board voltage transformers are harnessed for metering protection, power measurement, and catenary voltage measurement, playing an essential role in the entire high-voltage system. EMUs obtain data of the grid voltage and current through the electromagnetic voltage transformer to analyze

---

Content from this work may be used under the terms of the Creative Commons Attribution 3.0 licence. Any further distribution of this work must maintain attribution to the author(s) and the title of the work, journal citation and DOI.

Published under licence by IOP Publishing Ltd

their power qualities, including the basic information of traction network energy, such as power factor, harmonic, active power, reactive power, apparent power, etc., and analysis results of power qualities, such as voltage deviation, DC deviation, voltage fluctuation, and waveform distortion and so on.

The online detection scheme of instantaneous voltage has been studied both at home and abroad. The main detection scheme in China is to use a small number of coils in parallel at the end of the primary side winding as the parallel tap voltage. The parallel tap voltage at the primary side is always linearly related to the primary side voltage, thus realizing the voltage detection.[4] Foreign countries propose a high-frequency black box model of power transformers, which is based on the frequency-related voltage transfer function determined by the application of the four-terminal network theory.[5-6] In the traditional online devices for instantaneous over-voltage detection of EMUs, the method of adding non-contact probes is adopted to obtain detailed network voltage data. However, this method requires additional equipment. In order to achieve the measurement of signals in different frequency bands, multiple probes are in need. Besides, those probes themselves are vulnerable to electromagnetic interference, which affects the measurement accuracy.[7] Therefore, this paper utilizes the existing voltage transformer of EMUs to reconstruct the primary voltage waveform by obtaining the secondary side voltage waveform of the transformer, so as to monitor its transient over-voltage.

## 2. Principle in Reconstruction of Transformer Voltage Waveform

For the power frequency transformer, when there is an over-voltage signal $U_1(t)$ on the high-voltage bus passing through the voltage transformer, the measurable low-voltage side voltage $U_2(t)$ can be obtained. Ideally, the correspondence is available by the turns ratio and the internal admittance of the transformer. However, under high-frequency conditions, the internal characteristic admittance of the transformer follows the changes in frequency. Therefore, by virtue of the wide-band transmission characteristics of the transformer, an inverse calculation model can be established, and the function of the corresponding characteristic impedance of the transformer at different frequencies can be gained, so that the over-voltage full waveform $U_1(t)$ on the high-voltage port, i.e. the high-voltage bus, can be inversely deduced. The diagram of the transformer is displayed in Figure.1.

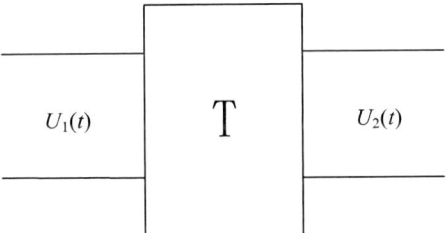

Fig.1 Transformer Diagram

$T$ is the equivalent port output parameter of a voltage transformer (TV), and the frequency-domain voltage signals obtained by the Fourier transform of $U_1(t)$ and $U_2(t)$ are $U_1(S)$ and $U_2(S)$ respectively. Then Formula (1) can be obtained:

$$\begin{bmatrix} U_1(S) \\ I_1(S) \end{bmatrix} = T \begin{bmatrix} U_2(S) \\ -I_2(S) \end{bmatrix} \tag{1}$$

In this Formula (1) $T = \begin{bmatrix} A(S) & B(S) \\ C(S) & D(S) \end{bmatrix}$, since the secondary side's measuring end of the voltage divider has a large impedance, it can be seen as an open circuit with a small current. Thus, if $I_2(S)=0$, the original formula is Formula (2):

$$U_1(S) = A(S)U_2(S) \tag{2}$$

$U_2(S)$ is a known measuring end, and $A(S)$ is a function of the characteristic admittance of voltage transformers at different frequencies. Targeting the problems of voltage with harmonics, that is, different or high-frequency waves, if the function $A(S)$ can be obtained, then it can be convolved with the known quantity $U_2(S)$ to calculate the corresponding primary side voltage $U_1(S)$ at each frequency in the harmonics, and then the time-domain $U_1(t)$ can be acquired through inverse Fourier transform.

If a sweep frequency instrument, the spectrum analyzer, is available, it can be employed to obtain the gain or attenuation curve of the transformer in a certain frequency range[8], as shown in Figure.2:

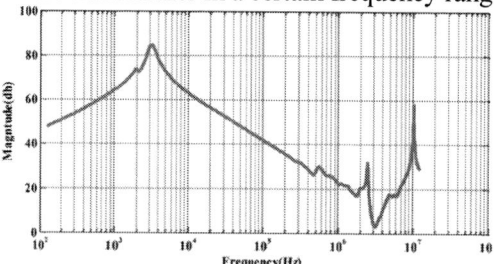

Fig.2 Curve of the Transformer at a Certain Frequency

The above figures present the frequency gain curve of a certain transformer measured by the sweep frequency method in the frequency range from 100 Hz to 10,000 kHz. A number of points are taken from the curve for vector fitting, and the function $T$ can be acquired. In Figure.3, the comparison between the fitting results and the sweeping curve is illustrated, and the error is slight. Therefore, the transfer function $T$ can be directly obtained and convolved with the voltage curve.

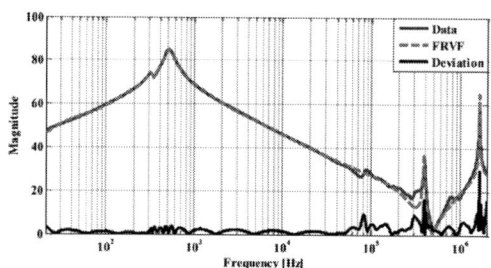

Fig.3 Comparison of Sweep Results and Fitting Results

## 3. Simulation Experiment

### 3.1 Circuit Model of Transformer

In Figure.4, the circuit model of the electromagnetic transformer is illustrated, wherein $R_1$ is the primary winding resistance; $L_1$ denotes the primary leakage inductance; $R_2$ means the secondary winding resistance; $L_2$ refers to the secondary leakage inductance; $L_m$ indicates the excitation inductance; $R_m$ stands for the excitation resistance.

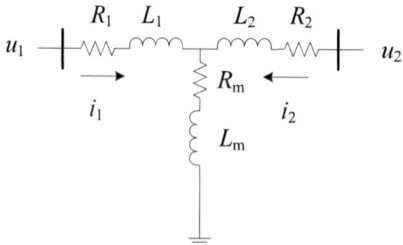

Fig.4 Circuit Mode of Electromagnetic Transformer

According to the transformer model, the circuit can be built in PSIM for the simulation experiment.

As shown in Figure.5, the primary side on the left is the primary side grid voltage, in which the common 3, 7, 25, 27, 47, and 99 harmonics in the grid are built, and the corresponding amplitude percentage is also given by the previous experimental results. The right side is the acceptable secondary side voltage on the train. The transformer model is set as 250: 1, the theoretical value of the peak in the primary voltage measurement is 25,000 V in the harmonic-free state, and the peak value of the secondary side is 100 V.

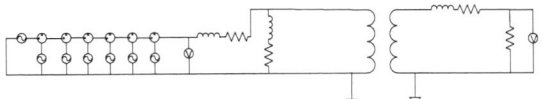

Fig.5 Simulation Circuit Model

It can be found that the secondary side voltage is as shown in Figure.6, and the peak voltage is about 100 V. The existence of harmonics calls forth slight oscillation.

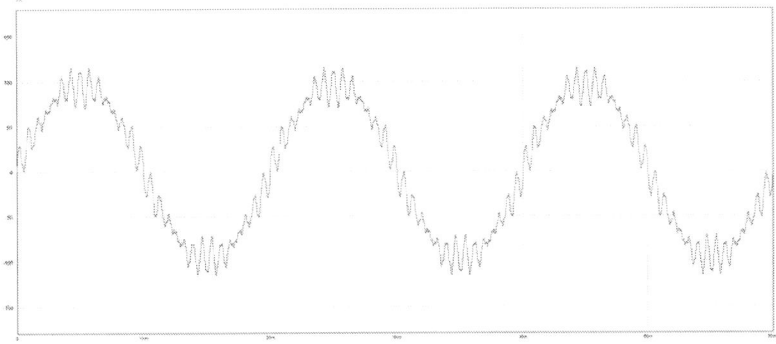

Fig.6 Secondary Side Voltage Waveform

*3.2 Frequency Sweep Simulation Experiment of Transformer*

A small signal model is established at first. As illustrated in Figure.7, a small signal with 5 Vpp and 50 Hz is introduced at the primary side to sweep the transformer model at 10Hz-5kHz. Since the harmonics of the grid voltage are mainly generated within 100 times, 5 kHz in the grid is the maximum frequency required for the sweep.

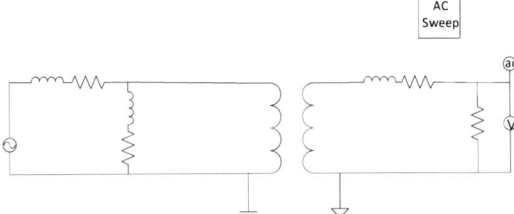

Fig.7 Small Signal Model of Simulation Experiment

The sweep results are presented in Figure.8, which are the Bode plots of amplitude, phase, and frequency respectively. By virtue of the calculation, the Bode plot gain of the transformer at 250: 1 should be -48 dB. The experimental results meet the requirements, and the transformer tends to decay with the increase of frequency, which conforms to the characteristics of a power frequency transformer.

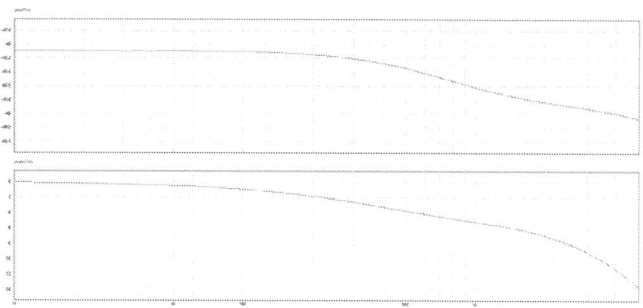

Fig.8 Sweep Frequency Curves of Simulation Experiment

### 3.3 Calculation Result

First, the corresponding Bode plot gain at each frequency is available in view of the sweep results, and then the frequency-domain waveform can be acquired by conducting FFT transformation on the secondary side time-domain waveform, as shown in Figure 9.

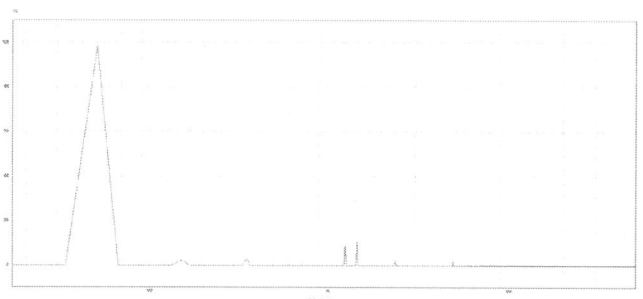

Fig.9 Frequency-Domain Waveform of Secondary Side Voltage

On the basis of observation, except at the power frequency of 50 Hz, the frequency-domain waveform also has the corresponding amplitude at a specific frequency due to the existence of harmonics. In Table.1, data is tabulated.

Tab.1 Corresponding Amplitude Gain at Different Frequencies

| Time of harmonic | Harmonic percentage | Frequency (Hz) | Gain (dB) | Primary side amplitude (V) | Secondary side amplitude (V) |
|---|---|---|---|---|---|
| 1 | 100% | 50 | -48.09 | 25000 | 98.11 |
| 3 | 2% | 150 | -48.12 | 500 | 1.95 |
| 7 | 3% | 350 | -48.24 | 750 | 2.89 |
| 25 | 10% | 1250 | -48.69 | 25000 | 9.18 |
| 29 | 12% | 1450 | -48.74 | 3000 | 10.84 |
| 47 | 2% | 2350 | -48.86 | 500 | 1.8 |
| 99 | 2.4% | 4950 | -49.06 | 600 | 2.09 |

In Table.1, there are 6 kinds of harmonics in the grid voltage, namely 3, 7, 25, 29, 47, and 99 times, and the harmonic percentages are 2%, 3%, 10%, 12%, 2%, and 2.4% respectively. The frequency-domain waveform diagram corresponds to 150 Hz, 350 Hz, 1250 Hz, 1450 Hz, 2350 Hz, and 4950 Hz, as well as the relevant secondary side amplitudes. By calculating them with different gains in the Bode plot, the corresponding primary side amplitude at the same frequency can be obtained, as shown in Table.2.

Tab.2 Corresponding Primary Side Voltage Amplitude at Different Frequencies

| Primary side amplitude (V) | Secondary side amplitude (V) | Gain (dB) | Calculation results of the primary side (V) |
|---|---|---|---|
| 25000 | 98.5 | 0.00394 | 25000 |
| 500 | 1.96 | 0.003926 | 499 |
| 750 | 2.9 | 0.003873 | 740 |
| 25000 | 9.19 | 0.003677 | 2499 |
| 3000 | 10.97 | 0.003656 | 3001 |
| 500 | 1.8 | 0.003606 | 499 |
| 600 | 2.11 | 0.003524 | 599 |

By observing the amplitude measured at the primary side (first column on the left) and that one through calculation of the secondary side and gain (first column on the right), it can be seen that the error is almost 0%. Therefore, the reliability and accuracy of the simulation experiment, that is, the algorithm, are verified. In Figure.10, the calculated result is displayed as the frequency-domain waveform of the primary side.

Fig.10 Frequency-Domain Waveform of Primary Side Voltage

Finally, the time-domain waveform of the primary side voltage is generated through IFFT transformation, as illustrated in Figure.11. It is exactly the same as the primary side waveform of the constructed grid voltage, with a peak value of 25,000 V, producing small fluctuations because of harmonics.

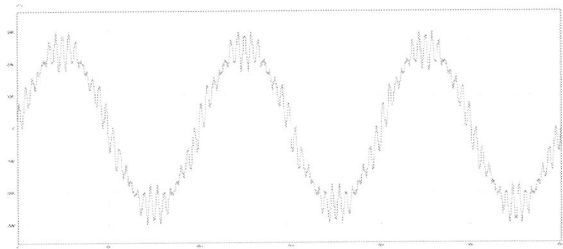

Fig.11 Primary Side Voltage Waveform

## 4. Frequency Sweep Experiment of Laboratory Transformer

### 4.1 Experimental Instrument

In this experiment, the frequency spectrum characteristic analyzer PSM3750 (with signal transmitting function at 5 Vpp and 50 Hz as well as a two-way signal receiving function of CH1 and CH2) of British Newtons4th Ltd is utilized for the frequency sweep experiment. The secondary side resistance is 7.5 ohms, and the transformer measured is a CR350 voltage transformer (25000: 100:100, voltage: 100 V, current: 200 mA). The physical diagram of the onboard transformer is displayed in Figure.12.

Fig.12 Image of Transformer

## 4.2 Principle of Frequency Sweep Experiment

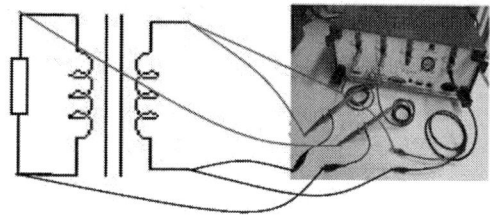

Fig.13 Schematic Diagram of Frequency Sweep Experiment

The wiring of the frequency sweep experiment is the same as that of the simulation experiment. As demonstrated in Figure.13, CH2 is connected to both ends of the secondary side resistance of the transformer. Besides, CH1 and the signal transmitting end are connected to both ends of the primary side of the transformer. The small signals are output by the instrument, with a frequency sweep ranging from 10 Hz to 5 kHz, and the spectrum analysis diagram of CH2/CH1 can be accessed.

## 4.3 Experiment Result

The Bode plots obtained by the frequency sweep ranging from 10 Hz to 5 kHz, that is, the curve of gain-frequency as well as the curve of phase-frequency, are displayed in Figures.14 and15. It can be observed that the waveform is basically consistent with the simulation experiment effect, and it is basically maintained at about - 50dB in the low-frequency range.

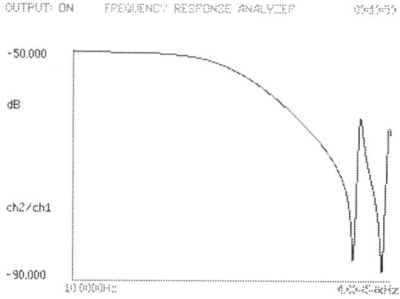

Fig.14 Amplitude-Frequency characteristic Bode Plot in Log Mode

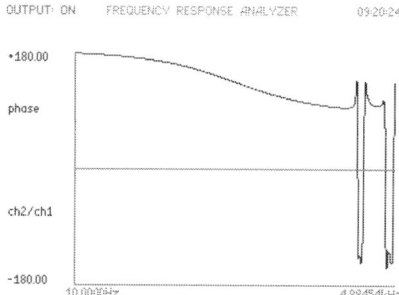

Fig.15 Bode Plot of Phase-Frequency Characteristics in Log Mode

It can be figured out from Figure.14 and Figure.15 above that when the frequency reaches about 2.3 kHz, the Bode plot starts to oscillate, which is because the parasitic parameters inside the transformer cause the attenuation to rebound. The waveform has poles and zeros and oscillates between - 60 dB and - 80 dB, which will reduce the impact of the transformer attenuation on the waveform at the secondary side which is not similar to that of the primary side. The overall curve is in line with the characteristics of the power frequency transformer.

The log image is converted into a linear mode to get the experimental results in Figure.16 and Figure.17 below:

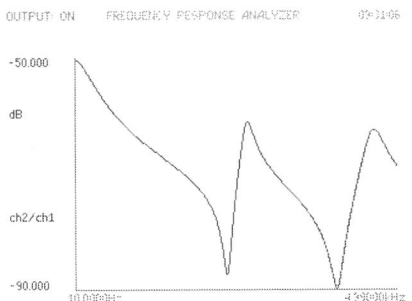

Fig.16 Amplitude-Frequency Characteristics Bode Plot in Linear Mode

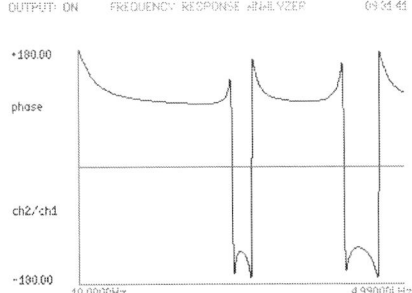

Fig.17 Phase-Frequency Characteristics Bode Plot in Linear Mode

According to the linear curve, the fundamental frequency required by the algorithm and the gain and phase results under different specific harmonics can be obtained. Thus, based on the experimental data, it can be fitted into a function. Its image tends to oscillate at a high frequency. Every time it attenuates for a while, due to the influence of parasitic parameters, it will have a trend of regression to the ratio at power frequency, which is generally between - 50 dB and - 90 dB, so as to ensure that the harmonic amplitude at high frequency can be smoothly transmitted to the secondary side of the transformer.[9]

This scheme is adopted in the online monitoring equipment of power quality. In the algorithm, the time-domain voltage obtained from the sampling end is transformed by FFT to acquire the real-time frequency-domain voltage, that is, the amplitude and phase of each corresponding harmonic frequency are obtained, and then the time-domain voltage is convolved with the fitting function and converted by IFFT to obtain the time-domain waveform diagram of the grid voltage, which can be displayed on the screen to capture the over-voltage of the grid in real time. The results are presented in Figure.18 below, wherein, Figure (1) is the voltage waveform diagram measured at the primary side of the transformer; Figure (2) shows the voltage waveform diagram at the primary side obtained by the conversion algorithm of the secondary side voltage waveform; Figure (3) denotes the voltage waveform diagram at the primary side directly obtained by equal proportion calculation of the secondary side voltage waveform. By comparison, Figure (1) is similar to Figure (2) in terms of time and amplitude, while Figure (3) obviously has large amplitudes, which differ greatly from the measured values. Therefore, according to the experiment, the scheme of reconstructing the primary side waveform from the secondary side of the transformer has accuracy and reliability, which can be used as the application scheme of the online monitoring of instantaneous over-voltage intelligent monitoring of EMUs.

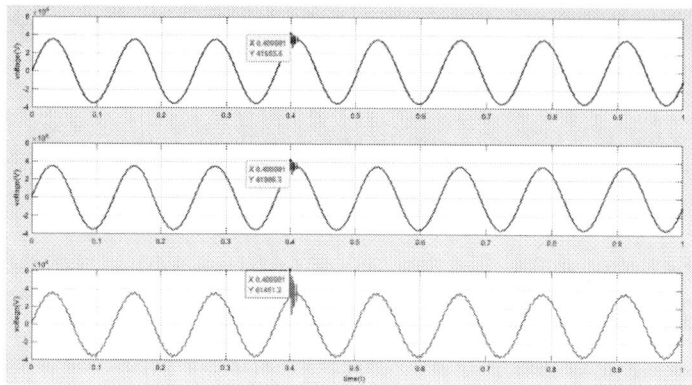

Fig.18 Comparison of Experimental Voltage Wave forms

## 5. Conclusion

This paper proposes a scheme that can utilize the secondary side of the transformer to reconstruct the voltage waveform of the primary side to monitor the transient over-voltage of the power grid. The reliability and accuracy are verified by simulation and experiment. The over-voltage signal is available from the junction cabinet of the voltage transformer in the vehicle. This signal is a low-voltage signal of hundreds of volts, which avoids the insulation problems caused by high-voltage signal acquisition. The monitoring system can be placed in the electrical cabinet in the vehicle to avoid the repercussions for train operation and passenger safety. Through the collection and analysis of instantaneous over-voltage signals, data can be provided for the accident analysis of onboard high-voltage equipment. By the statistical analysis of the amplitude, action length, and action times of the instantaneous over-voltage, the voltage value of the on-board high-voltage electrical equipment under the actual operating conditions can be acquired, which offers data support for the aging and lifespan research of the on-board high-voltage electrical equipment.

## Author profile

Zhongjiang Sun: Male; a Master born in 1990; major in power electronic technology, design of the power system for EMUs, etc.

Suyue Liu(Corresponding author): Female; a Master born in 1999; major in power electronic technology, design of the power system for EMUs, etc.

Zhiming Yang: Male; a Master born in 1990; major in power electronic technology, design of the power system for EMUs, etc.

Shuangfeng Xu: Female; a Master born in 1994; major in power electronic technology, design of the power system for EMUs, etc.

## Reference

[1] Zhang Z. Y., Huang T., Ren Y. Y. Y., et al. On-line Over-Voltage Monitoring Device with Wide Frequency Characteristics of Power Station Equipment[J]. High-Voltage Technology, 2011, 37(2):8.

[2] Liu B., Ren L., Xu X. H., et al. On-line Monitoring System of Grid Over-Voltage Based on Virtual Instrument[J]. Electric Power Automation Equipment, 2007, 027(002):97-100.

[3] Yan J. Research on the Numerical Inversion Calculation Methods of Over-Voltage Based on the Wide Band Transfer Characteristics of Apparatus[D]. North China Electric Power University (Baoding), 2008.

[4] Gao B. W., Jin J. On-line Over-Voltage Monitoring Method for On-Board Voltage Transformer of High-Speed EMUs[J]. The World of Inverters, 2019(8):90-94.

[5] D. Filipović-Grčić, B. Filipović-Grčić and I. Uglešić, "High-Frequency Model of the Power Transformer Based on Frequency-Response Measurements," in IEEE Transactions on Power

Delivery, vol. 30, no. 1, pp. 34-42, Feb. 2015, DOI: 10.1109/TPWRD.2014.2327061.

[6] Che Y., Wang Q., Li X., and Zhang J. "Active Reset System of Linear Transformer and Loss Analysis," 2021 IEEE 4th International Conference on Electronics Technology (ICET), 2021, pp. 533-537, DOI: 10.1109/ICET51757.2021.9450998.

[7] Lan H. T., SiMa W. X., Yao C. G., et al. A Study on Collection Methods in On-line Monitoring Data on Over-Voltage of High-voltage Grid[J]. High-Voltage Technology, 2007(3):79-82.

[8] Faraj M A, Mousa S K, Shuaieb W S, et al. Power Transformer Modelling Based on Vector Fitting Method. 2020.

[9] B. Gustavsen and A. Semlyen, "Application of vector fitting to state equation representation of transformers for simulation of electromagnetic transients," in IEEE Transactions on Power Delivery, vol. 13, no. 3, pp. 834-842, July 1998, Doi: 10.1109/61.686981.

# Rational Design of Future Potential Electric Aircraft

**Mingze Xi***

Tianjin Yinghua Experimental School, Tianjin, 301700, China

* Corresponding author: 2342077@tjyh2003.com

**ABSTRACT:** With the development of science and technology, the shortage of fossil energy has become a problem that cannot be ignored. The main significance of studying electric aircraft is to reduce energy consumption and achieve green travel. The electric airplane project is very unusual. It can change the energy use of the world. If the electric airplane can be designed, it will be used in other fields, such as aviation. The advantages of electric aircraft, such as low cost, high efficiency, and sustainability, can reduce the world's oil energy consumption. In addition, it is believed that many people have built electric vehicles. The noise of electric vehicles will be quieter than that of fossil fuel aircraft. This green energy can make the world develop and progress faster. Furthermore, the low price, energy conservation, environmental protection, and low cost have huge commercial potential. The paper organization mainly consists of the new battery, electric motor vs. jet, thermal energy, and required energy.

## 1. INTRODUCTION

Through the current technology, people can produce electric energy vehicles. However, whether this technology can be applied to aircraft is still a relatively big challenge, because it is difficult to find a battery with enough energy density, a reasonable engine, and a small mass to replace the current oil [1, 2]. However, science and technology will significantly progress if people succeed in solving this problem, then drive global development. First, its advantages are obvious in terms of lower cost, sustainability, fast response, and effectiveness. Also, electric aircraft can reduce direct carbon emissions and fuel costs by up to 90%, maintenance costs by up to 50%, and noise by 70% [3]. The main challenge people face is finding new batteries and electric engines with perfect performance objective factors in all aspects and calculating the exact data. Before this research on this topic in the Iceland industry, the "Rafmagnsflug" company staged a historical event and operated the nation's first electric plane, and the President and the Prime Minister were among the first customers of this electric aircraft [4]. This event aimed to find a new air energy exchange, letting people adapt to and accept the new technology and spread it across the country. The electric plane, which was built in Slovenia in 2021 and named TF-KWH, weighs 600 kg but has the disadvantage of being able to carry only one person at a time. The airline claims last week's service and the success of taking off represents the first step on an important journey towards greener aviation. Iceland's first electric plane is expected to allow public members to buy sightseeing flights and experience zero-emission aircraft [5]. Icelandair, like many other airlines around the world, aims to cut its carbon emissions per ton kilometer of operation in half by 2030 compared with 2019 levels in a sustainable manner. The company will also need new aircraft technology, continuous operational improvements, sustainable aviation materials, and new high-energy density batteries to achieve this target.

---

Content from this work may be used under the terms of the Creative Commons Attribution 3.0 licence. Any further distribution of this work must maintain attribution to the author(s) and the title of the work, journal citation and DOI.

Published under licence by IOP Publishing Ltd

## 2. PROBLEMS AND CHALLENGES OF CHOOSING A BATTERY

First, this paper aims to design an all-electric jumbo jet (Figure 1) [6]. To begin with, the battery is a necessary accessory. It is generally known that the battery's energy density is much lower than that of kerosene, so the battery will be very heavy. The aircraft needs more energy to fly.

### 2.1 Why Not the Lithium-ion Batteries

However, higher energy demand will increase the battery's weight, and a cycle will occur [7]. At present, Tesla cars use lithium-ion batteries, but the energy density of lithium-ion batteries is not enough to support aircraft takeoff, and therefore higher energy density is needed to be sought (Figure 2). Lithium-ion batteries are the cells of choice for flight because they pack the most power into the lowest mass of any conventional battery, but they have one disadvantage: they can short-circuit and fall into what is known as "thermal runaway," ending in combustion.

Figure 1. Image of a typical electric aircraft [6].

### 2.2 What Is a Fluoride Battery

To this end, the fluoride battery stands out among many batteries [7]. First, fluoride's service life is eight times that of lithium-ion batteries because it is an anion, making it difficult to stabilize. Nevertheless, a research group accidentally discovered a liquid point solute that can stabilize elements last year, solving a defect in fluoride batteries. However, the requirements for storage conditions of fluoride are relatively high, so it is hard to use for the time being. Hopefully, people can find the appropriate method to store it in the future [8] and the way that dynamic force cells measure the density. First, energy density is the amount of energy stored per unit of space or mass of matter. The energy density of a battery is the amount of electric energy released per unit volume or mass of the battery.

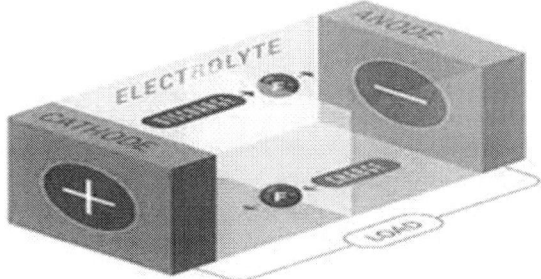

Figure 2. Scheme of lithium-ion battery components [8].

## 3. ELECTRIC MOTORS VS JET

In earlier aircraft designs, mechanical, hydraulic, and pneumatic transmission power came from the aircraft engine, which accounted for 95 percent of the power output for propulsion and about 5 percent for various control components and systems [9]. This has led to a dramatic increase in the number of liquid and gas lines in the aircraft, which has a measurable impact on the aircraft's total weight and fuel consumption, reducing the aircraft's effective load and producing more carbon emissions. At present, there are six mainstream electric aircraft solutions in the world. The performance of the aircraft influences its ability to reduce carbon dioxide emissions. The aircraft components' performance would influence carbon dioxide emissions, the aircraft configuration's design, and the mission area performed. These six technical solutions can be further divided into one fully electric drive solution, three electric-oil hybrid solutions, and two turbo-electric propulsion solutions.

After that, the plan is to search for a new method. Based on the survey and current society, there are three new energy sources to reduce carbon emissions: SAF sustainable aviation fuel, hydrogen fuel, and special gas. SAF sustainable aviation fuel is a new fuel made from animal waste oil. Compared with the previous aviation fuel, its biggest difference is that it is very clean and will greatly reduce greenhouse gas emissions during its use. Then, hydrogen fuel can completely replace kerosene used by engines and eliminate carbon dioxide emissions, reducing greenhouse gas emissions by 2/3. Then comes the pure electric propulsion system, which is also an energy source studied. It can also reduce some $CO_2$ emissions, but not all of them. Also, the gas turbine engine is a kind of engine that uses rotating parts to absorb kinetic energy from the fluid flowing through it. It is a kind of internal combustion, mainly divided into turbojet and turbofan. The compressor, combustion chamber, and turbine are essential parts of the three turbine engines. In the low carbon era, the advanced aerospace development in the future has put forward newer and higher requirements for the use of cost, speed, environmental protection, and fuel efficiency. The data in the table can help compare the three types of new energy. It can be seen that compared with traditional kerosene, sustainable air fuel has no requirements for the design of the aircraft itself, and SAF is not inferior to aviation kerosene in practical operation, while hydrogen fuel and pure electric engines need to change the structural design of aircraft and engines.

## 4. THERMAL ENERGY

The next part is about thermal energy, which is also a significant part, and there are mainly three steps and two issues. The first issue is the low external temperature. The temperature drops five degrees for every 10,000 meters in height, so the problem is that the solid electrolyte of fluoride-ion battery requires a high temperature, mainly 15 degrees, to function. In addition, the second problem is the thermal energy runaway, which includes overheating, combustion and explosion, and heat accumulation and gas release. In 2013, a fire incident occurred in the Boeing 787 Dreamliner of a well-known aircraft company because there are safety risks in the fire extinguishing system. In addition, the auxiliary unit fire overheating should also be blamed. There are two solutions. The first is to add Thick Separators. The second point is to evaluate FIB made of different types of electrodes. Moreover, people can also evaluate FIBs made from different types of electrodes and test out the thermal runaway point and evaluate the potential risks of FIBs made from different cathodes and anodes. Nevertheless, it should be noted that the cathode material BIF is harmful to human skin. When it comes to the three steps, they can solve those issues. The first step is to use BTFE. Researchers from Honda's research institute, Caltech, and JPL developed BTFE. While the room temperature is 20 degrees, it is still way higher than the -5 degrees. This kind of cold temperature results in an increase in the internal resistance of batteries, which lowers the capacity of batteries.

Furthermore, step two is to apply the heater. This method controls the temperature and ensures that the fluoride battery can work smoothly in the aircraft. Finally, the last step, which is also the third step, is adding a backup energy source. Solar panels can be added to the plane to harness the sun's energy and avoid accidents if fluoride batteries do not function well. For example, a long-range solar plane in Switzerland uses this principle.

## 5. ENERGY REQUIRED

The energy requirement is the last issue that needs to be discussed. This very serious issue needs much attention because it includes many calculations. First, the question is why an airplane can stay in the air. According to Newton's third law, the downward thrust of the wing produces an equal and opposite upward force. Since the wings are so thin, the height difference can be ignored. Supposing that the pressure on the upper body of the aircraft is 400 ml/h, and the downward pressure is 300 ml/h [10]. Then, "Bernoull's Equation" needs to be used:

$$P_1 + \frac{1}{2}PV^2 + pgy_1 = P_2 + \frac{1}{2}PV_2^2 + pgy_2 \qquad (1)$$

where y represents height, P represents pressure, and it can be concluded that $P2 > P1$, so the upward force is greater, and the aircraft could fly up (Figure 3).

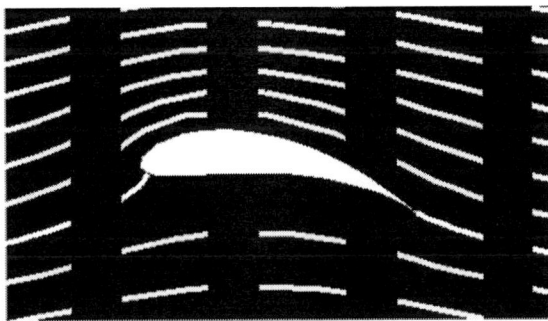

Figure 3. The picture of the flow of wind speed.

*5.1 The Total Energy of an Airplane in the Air*

Then the next step is to calculate the total energy of an airplane in the air. Assuming that the maximum take of weight is 7500 kg, and cruise height is approximately 10000 m. The cruise speed is 128.62 m/s, and the gravity of Earth is 9.8 N/kg. And putting all the messages in the Formula:

$$E = \frac{1}{2}mv^2 + mgh \qquad (2)$$

And it can be reached that the total energy is $7.98 \times 10^8$ Joule. However, the engines mounted on the wings of the plane are required to provide additional energy per time and power to keep the aircraft flying at a constant altitude and speed. Also, the next step is to calculate the power required for the airplane to stay in the air. The Formula is:

$$P = (dh/dt)mg \qquad (3)$$

(dh/dt) is the decent speed required to keep the plane's speed constant if the engines fall. The glide ratio of the plane must be 25/1. The glide ratio is how far it travels horizontally compared to how far it drops when the engines fail. Finally, an Equation could be drawn as below:

$$v \times \frac{1}{\text{glide ratio}} \qquad (4)$$

*5.2 The Lift Equation*

On the other hand, another question needs to be studied: how the airplane takes off by the lift equation. The basic requirement of taking is that the lift needs to be greater than the weight pulling the plane down. A new Equation is used:

$$\text{Lift} = \frac{2}{\text{coefficient} \times \text{density} \times \text{velocity}^2 \times \text{wing area}} \qquad (5)$$

The coefficient contains all the complex dependencies and is usually determined experimentally. One diagram illustrates that if the angle is attacked from -5 degrees to 15 degrees, the lift coefficient increases, while after the angle attack reaches 15 degrees, the coefficient will decrease. The greater the lift coefficient, the greater the lift. The angle attack is the angle between land and the nose of an aircraft. During the flight, the wings are parallel to the sky, while the wings of the plane should be perpendicular to the ground when taking off, so the surface of the wings can feel more pressure. After all, the final answer regarding the energy requirement of each plane can be calculated.

## 6. CONCLUSION

To conclude, the sodium-ion battery has its advantages and specific applications compared with the lithium-ion battery. The current challenges of sodium-ion batteries, the possible solutions and the prospect of future development direction are areas waiting to be explored further: looking for cathode materials with high energy density and functional density; looking for negative electrode materials with small volume change during the cycle are the area that people should concentrate on. Besides, all the issues arising in the designing process must be resolved, and all the information needed should be calculated. In the future, if people can have as many electric planes as people have electric cars now, then the whole world will change qualitatively, and at the current rate of technological development, the breakthrough is bound to happen soon.

## REFERENCES

[1] Epstein, A.H. and O'Flarity, S.M., "Considerations for reducing aviation's $CO_2$ with aircraft electric propulsion," Journal of Propulsion and Power, 35(3), 572-582 (2019).

[2] Patterson. T., Dreamliner battery probe ends: 8 questions and answers, 2014, https://www.cnn.com/travel/article/boeing-787-dreamliner-investigation-report/indexhtml

[3] Sripad, S., Bills, A. and Viswanathan, V., "A review of safety considerations for batteries in aircraft with electric propulsion," MRS Bulletin, 46(5), 435-442 (2021).

[4] Tariq, M., Maswood, A.I., Gajanayake, C.J. and Gupta, A.K., "Aircraft batteries: current trend towards more electric aircraft," IET Electrical Systems in Transportation, 7(2), 93-103 (2017).

[5] Catwell (2018, December 17), Fluoride battery to replace lithium. Blog-Power&Energy-element14 Community. Retrieved August 20, 2022, from https://community.element14.com/technologies/power-management/b/blog/posts/fluoride-battery-to-replace-lithium

[6] Sarlioglu, B. and Morris, C.T., "More electric aircraft: Review, challenges, and opportunities for commercial transport aircraft." IEEE transactions on Transportation Electrification 1(1), 54-64 (2015).

[7] Campbell, M. (2022, February 9), Could replacing lithium with salt solve the battery crisis? Euronews. Retrieved August 21, 2022, from https://www.euronews.com/green/2022/02/09/we-re-facing-a-lithium-battery-crisis-what-are-the-alternatives

[8] Fluoride-ion breakthrough promises 10x energy density compared with lithium-ions-news. EEPower. (n.d.). Retrieved August 21, 2022, from https://eepower.com/news/opposites-in-nature-fluoride-and-lithium-compete-for-higher-energy-batteries/#

[9] Wheeler, P. and Bozhko, S., "The more electric aircraft: Technology and challenges," IEEE Electrification Magazine, 2(4), 6-12 (2014).

[10] Tahir, M., Khan, S.A., Khan, T., Waseem, M., Khan, D. and Annuk, A., "More electric aircraft challenges: A study on 270 V/90 V interleaved bidirectional DC-DC converter," Energy Reports, 8, 1133-1140 (2022).

# Adaptive neural control for chaotic permanent magnet synchronous motor with asymmetric input saturation

**Hongshan Liu[1], Huibo Liu[1*], Yanwei Zhao[2]**

[1] School of Information Engineering, Inner Mongolia University of Science and Technology, Baotou014017, China.

[2] School of Control Science and Engineering, Bohai University, Jinzhou121013, China.

**Abstract.** This paper develops an adaptive neural control scheme for chaotic permanent magnetic motor with asymmetric input saturation and full-state constraints. To solve the state constrain problem, a barrier Lyapunov function is introduced. Then, a sliding-mode differentiator is designed to avoid the "explosion of complexity" issue. A simulation example is given to verify the validity of the proposed scheme.

## 1. Introduction

Permanent magnet synchronous motors (PMSM) are of great significance in industrial applications owing to their high efficiency, high speed, high power density, and high torque to inertia ratio. However, the stability problems of the PMSM in industrial operation is still a challenging task. This is mainly because that its dynamic model is multivariable, nonlinear, and even chaotic attractors with systemic parameters. To cope with the challenging task, many scholars have devoted to develop nonlinear control methods for PMSM and many achievements have been proposed [1,3]. The chaotic behavior in PMSM is often encountered, which may lead to system instability even cause system collapsed. Recently, backstepping-based adaptive control technique has been proposed and several excellent achievements for chaotic PMSM has been obtained [2,4]. However, a common challenge of the above control schemes is that the "explosion of complexity" problem existed in the conventional backstepping methodology is not addressed. In order to eliminate this issue, a novel first-order sliding-mode differentiator is proposed to generate the derivative of virtual control signal, such that the computation is greatly simplified.

On the other hand, constraints are a common phenomenon that extensively exist in motor systems, which may bring negative effects on system behaviors. Thus, researches on constrained motor systems have a practical significance. Unfortunately, for the class of chaotic PMSM systems, the problem of adaptive neural control has not been adequately studied, which promotes our works.

Inspired by the above analysis, this paper studies an adaptive neural control of chaotic PMSM subject to input and state constrains. A slide-mode differentiator is constructed such that the computation burden of the proposed scheme is reduced. Compared with the adaptive control scheme of constrained PMSM [4], an asymmetric saturation nonlinearity is considered in this paper to enhance the practicability.

## 2. Problem formulation

The dynamic of the chaotic permanent magnet synchronous motor (PMSM) is described as

$$\frac{d\Theta}{dt} = w_s$$

$$\frac{dw_s}{dt} = \bar{\sigma}(i_q - w_s) - T_L$$

$$\frac{di_q}{dt} = -i_q - i_d w_s + \bar{\gamma} w_s + u_q \tag{1}$$

$$\frac{di_d}{dt} = -i_d + i_q w_s + u_d$$

where $i_d$ and $i_q$ represent the axis currents, $\Theta$ stands for the rotor position, $T_L$ denotes the load torque, $w_s$ is the rotor angular, $u_q$ and $u_d$ act as the voltages velocity, and $\bar{\sigma}$ and $\bar{\gamma}$ represent the positive parameters. Defining $x_1 = \Theta$, $x_2 = w_s$, $x_3 = i_q$, $x_4 = i_d$, the model of PMSM is rewritten as

$$\dot{x}_1 = x_2$$

$$\dot{x}_2 = \sigma(x_3 - x_2) - T_L$$

$$\dot{x}_3 = -x_3 - x_2 x_4 + \gamma x_2 + u_q(v_q) \tag{2}$$

$$\dot{x}_4 = -x_4 + x_2 x_3 + u_d(v_d)$$

Lemma 1: For non-negative constants $k_b i$ and errors $z_i (i = 1, \cdots, 4)$, there exists the inequality as $\log\left(k_{bi}^2 / \left(k_{bi}^2 - z_i^2\right)\right) \le z_i^2 / \left(k_{bi}^2 - z_i^2\right)$.

The objective of this paper aims to develop an adaptive neural control scheme for chaotic PMSM, such that the rotor position $x_1$ can track the desired position $x_d$. Suppose that the induction motor input $u(v)$ suffers from asymmetric saturation nonlinearity, which is given by

$$u(v) = sat(v) = \begin{cases} u_M, v > u_M \\ v, u_m \le v \le u_M \\ u_m, v < u_m \end{cases} \tag{3}$$

where $v(t)$ denotes the control input, $u_M$ and $u_m$ are the bounds of the motor input $u(v)$. Then, an asymmetric model is given to approximate the saturation nonlinearity, which is constructed as $\kappa(v) = u_b \times erf\left(\sqrt{\pi} v / 2u_b\right)$ with $u_b = (u_M + u_m)/2 + (u_M - u_m)/2$ and $erf(x) = \frac{2}{\sqrt{\pi}} \int_0^x e^{-t^2} dt$. To simplify the design progress, the model is rewritten as $u(v) = \kappa(v) + \Lambda(v)$. By introducing the mean value theorem, we conclude the equation as $\kappa(v) = \kappa(v_0) + \kappa_{v_a}(v - v_0)$ with $\kappa_{v_a} = \partial \kappa(v) / \partial \kappa |_{v=v_a} = \exp(-(\sqrt{\pi} v / 2u_b)^2) |_{v=v_a}$ and $v_a = av + (1-a)v_0, (0 < a < 1)$. By choosing $v_0 = 0$, the model is redefined as

$$u(v) = \kappa_{v_a} v + \Lambda(v), (0 < \kappa_{v_a} < 1) \tag{4}$$

where $\Lambda(v) = \kappa_{v_a} v - u(v)$ represents the bounded function.

## 3. Adaptive neural controller design and stability analysis

In this section, the adaptive neural controllers are designed for PMSM. The coordinate transformation is constructed as

$$z_1 = x_1 - x_d$$

$$z_2 = x_2 - \alpha_1$$

$$z_3 = x_3 - \alpha_2 \tag{5}$$

$$z_4 = x_4$$

where $z_i (i=1,\cdots,4)$ represent the errors, which are constrained by constants $k_{bi}$, $x_d$ denotes the desired signal, and $\alpha_1$ and $\alpha_2$ are the virtual controllers.

Step 1: Consider the lyapunov function as $V_1 = \frac{1}{2}\log\frac{k_{b1}^2}{k_{b1}^2 - z_1^2}$. In light of the coordinate transformation Eq. (5), the time derivative of $V_1$ is derived as

$$\dot{V}_1 = \vartheta_{z_1}(z_2 + \alpha_1 - \dot{x}_d) \tag{6}$$

where $\vartheta_{z_1} = z_1/(k_{b1}^2 - z_1^2)$. The virtual controller is constructed as

$$\alpha_1 = -r_1 z_1 - \frac{1}{2}\vartheta_{z_1} + \dot{x}_d \tag{7}$$

where $r_1$ is a positive parameter. Substituting Eq. (7) into Eq. (6), one has

$$\dot{V}_1 \leq -r_1 \vartheta_{z_1} z_1 + \frac{1}{2}z_2^2 \tag{8}$$

where the Young's inequality $\vartheta_{z_1} z_2 \leq \frac{1}{2}\vartheta_{z_1}^2 + \frac{1}{2}z_2^2$ has been used.

Step2: Choose the lyapunov function candidate as $V_2 = V_1 + \frac{1}{2}\log\frac{k_{b2}^2}{k_{b2}^2 - z_2^2} + \frac{1}{2}\tilde{W}_2^T\tilde{W}_2$ with $\tilde{W}_2 = W_2^* - \hat{W}_2$ being the error of $W_2^*$. Then, by virtue
of Eq. (5) and Eq. (8), the derivative of $V_2$ is described as

$$\dot{V}_2 \leq -r_1 \vartheta_{z_1} z_1 + \frac{1}{2}z_2^2 + \frac{z_2}{k_{b2}^2 - z_2^2}(\sigma(x_3 - x_2) - T_L - \eta_{20} - \rho_1) - \tilde{W}_2^T\dot{\hat{W}}_2 \tag{9}$$

where $\vartheta_{z_2} = z_2/(k_{b2}^2 - z_2^2)$ and $f_2(\chi_2) = -\sigma x_2 + z_2^2/2\vartheta_{z_2}$ with $\chi_2 = [x_d, \dot{x}_d, x_1, x_2]^T$. By virtue of the universal approximation ability of neural networks (NNs), the unknown function $f_2(\chi_2)$ is represented as $f_2(\chi_2) = W_2^{*^T}\varphi_2(\chi_2) + \varepsilon_2(\chi_2)$ with $\varepsilon_2(\cdot)$ and $\varphi_2(\cdot)$ being the bounded approximation error and Gaussian function.

In order to avoid the complicated calculation of $\dot{\ }\alpha 1$, the first-order slide-mode differentiator is introduced as

$$
\begin{aligned}
\dot{P}_{20} &= \eta_{20} \\
&= P_{21} - \xi_{20}\,|\,P_{20} - \alpha_1\,|^{(1/2)}\,sign(P_{20} - \alpha_1) \\
\dot{P}_{21} &= -\xi_{21}sign(P_{21} - \eta_{20}) \\
\dot{\alpha}_1 &= \eta_{20} + \rho_1, |\,\rho_2\,|\leq \bar{\rho}
\end{aligned} \tag{10}
$$

where $\dot{P}_{20}$, $\eta_{20}$ and $\dot{P}_{21}$ stand for the states of the differentiator, $\xi_{20}$ and $\xi_{21}$ are design parameters, and $\bar{\rho}$ represents a positive constant.

Then, the inequality Eq. (9) is rewritten as

$$
\begin{aligned}
\dot{V}_2 \leq -r_1 \vartheta_{z_1} z_1 + \vartheta_{z_2}(\sigma z_3 + \sigma\alpha_2 + W_2^{*^T}\varphi_2(\chi_2) + \varepsilon_2(\chi_2) - \\
T_L - \eta_{20} - \rho_1) - \tilde{W}_2^T\dot{\hat{W}}_2
\end{aligned} \tag{11}
$$

Note that the external load torque $T_L$ is unknown but bounded by a positive constant $T_M$, that is, $\|T_L\| \leq T_M$. With the help of Young's inequality, one has

$$\vartheta_{z_2}(\bar{\sigma}z_3 - T_L - \rho_1 + \varepsilon_2(\chi_2)) \leq 2\vartheta_{z_2}^2 + \frac{\bar{\sigma}^2 z_3^2}{2} + \frac{T_M^2}{2} + \frac{\bar{\rho}^2}{2} + \frac{\varepsilon_M^2}{2} \tag{12}$$

Where $\varepsilon_M$ is a positive constant which are the upper bound of approximation error $\varepsilon_2(\chi_2)$. Design the virtual controller and adaptive law as

$$\alpha_2 = \frac{1}{\bar{\sigma}}(-r_2 z_2 - 2\vartheta_{z_2} - \hat{W}_2^T \varphi_2(\chi_2) + \eta_{20})\qquad(13)$$

$$\dot{\hat{W}}_2 = \vartheta_{z_2} z_2 - d_2 \hat{W}_2\qquad(14)$$

where $r$ and $d_2$ denote the design parameters. Inserting the virtual controller Eq. (13), adaptive law Eq. (14) and inequality Eq. (12) into Eq. (11), we conclude

$$\dot{V}_2 \leq -\sum_{i=1}^{2} r_i \vartheta_{z_i} z_i + \frac{\bar{\sigma}^2 z_3^2}{2} + d_2 \tilde{W}_2^T \hat{W}_2 + \frac{1}{2}(T_M^2 + \bar{\rho}^2 + \varepsilon_M^2)\qquad(15)$$

Step 3: Define the lyapunov function as $V_3 = V_2 + \frac{1}{2}\log(\frac{k_{b3}^2}{k_{b3}^2 - z_3^2}) + \frac{1}{2}\tilde{W}_2^T \tilde{W}_2$. According to Eq. (5) and Eq. (15), the derivative of $V_3$ is described as

$$
\begin{aligned}
\dot{V}_3 \leq &-\sum_{i=1}^{2} r_i \vartheta_{z_i} z_i + \frac{\bar{\sigma}^2 z_3^2}{2} + d_2 \tilde{W}_2^T \hat{W}_2 - \tilde{W}_2^T \dot{\hat{W}}_2 + \frac{1}{2}(T_M^2 + \bar{\rho}^2 + \varepsilon_M^2) \\
&+ \frac{z_3}{k_{b3}^2 - z_3^2}(-x_3 - x_2 x_4 + \bar{\gamma} x_2 + u(v_q) - \dot{\alpha}_2)
\end{aligned}\qquad(16)
$$

Define $\vartheta_{z_3} = z_3 / (k_{b3}^2 - z_3^2)$ and $f_3(\chi_3) = -x_3 - x_2 x_4 + \gamma x_2 + \sigma^2 z_3^2 / 2\vartheta_{z_3}$ with $\chi_3 = [x_d, \dot{x}_d, x_1, x_2, x_3, x_4]^T$. The NNs are used to approximate the unknown function $f_3(\chi_3)$ as $f_3(\chi_3) = W_3^{*T} \varphi_3(\chi_3) + \varepsilon_3(\chi_3)$. The meaning of symbols $\varphi_3(\cdot)$ and $\varepsilon_3(\cdot)$ are same as step 2. Meanwhile, the first-order sliding-mode differentiator is introduced to represent $\dot{\alpha}_2$ as $\dot{\alpha}_2 = \eta_{30} + \rho_2$. Then, by virtue of the above results and saturation model Eq. (4), the inequality Eq. (16) is rewritten as

$$
\begin{aligned}
\dot{V}_3 \leq &-\sum_{i=1}^{2} r_i \vartheta_{z_i} z_i + d_2 \tilde{W}_2^T \hat{W}_2 - \tilde{W}_2^T \dot{\hat{W}}_2 + N_2 \\
&+ \vartheta_{z_3}(\mu_{v_a} v_q + d(v_q) + W_3^{*T} \varphi_3(\chi_3) + \varepsilon_3(\chi_3) - \eta_{30} - \rho_2)
\end{aligned}\qquad(17)
$$

By using Young's inequality, we can conclude

$$\vartheta_{z_3}(\mu_{v_a} v_q + d(v_q) - \rho_2 + \varepsilon(\chi_3)) \leq \frac{3}{2}\vartheta_{z_3}^2 + \vartheta_{z_3} v_q + \frac{1}{2}d_M^2 + \frac{1}{2}\bar{\rho}^2 + \frac{1}{2}\varepsilon_M^2\qquad(18)$$

where $d_m$ denotes a positive constant satisfied $\| d(v_q) \| \leq d_M$. Design the controller and adaptive law as

$$v_q = -r_3 z_3 - \vartheta_{z_3} - \hat{W}_3^T \varphi_3(\chi_3) + \eta_{30}\qquad(19)$$

$$\dot{\hat{W}}_3 = \vartheta_{z_3} \varphi_3(\chi_3) - d_3 \hat{W}_3\qquad(20)$$

where $r_3$ and $d_3$ represent the design parameters. Substituting Eqs. (19), (20) and Eq. (18) into Eq. (17), the inequality Eq. (17) is redescribed as

$$\dot{V}_3 \leq -\sum_{i=1}^{3} r_i \vartheta_{z_i} z_i + \sum_{i=2}^{3} d_i \tilde{W}_i^T \hat{W}_i + \frac{1}{2}\left(T_M^2 + 2\bar{\rho}^2 + 2\varepsilon_M^2 + d_M^2\right)\qquad(21)$$

Step 4: We construct the lyapunov function as $V_4 = V_3 + \frac{1}{2}\log(\frac{k_{b4}^2}{k_{b4}^2 - z_4^2}) + \frac{1}{2}\tilde{W}_4^T\tilde{W}_4$ .In light of Eqs. (4), (5) and (21), the time derivative of $V_4$ is derived as

$$\dot{V}_4 \le -\sum_{i=1}^{3} r_i \vartheta_{z_i} z_i + \sum_{i=2}^{3} d_i \tilde{W}_i^T \hat{W}_i - \tilde{W}_4^T \dot{\hat{W}}_4 + \frac{1}{2}\left(T_M^2 + 2\bar{\rho}^2 + 2\varepsilon_M^2 + d_M^2\right)$$
$$+ \frac{z_4}{k_{b4}^2 - z_4^2}(-x_4 + x_2 x_3 + \mu_{v_a} v_d + d(v_d)) \tag{22}$$

where $f_4(\chi_4) = -x_4 + x_2 x_3$ with $\chi_4 = [x_2, x_3, x_4]^T$ . According to the approximation ability of NNs, we have $f_4(\chi_4) = W_4^{*T}\varphi_4(\chi_4) + \varepsilon_4(x_4)$ with $\varepsilon_4(\cdot)$ and $\varphi_4(\cdot)$ being the same meaning as step 2. Then, defining $\vartheta_{z_4} = z_4 / (k_{b4}^2 - z_4^2)$ , the inequality Eq. (22) is rewritten as

$$\dot{V}_4 \le -\sum_{i=1}^{3} r_i \vartheta_{z_i} z_i + \sum_{i=2}^{3} d_i \tilde{W}_i^T \hat{W}_i - \tilde{W}_4^T \dot{\hat{W}}_4 + \frac{1}{2}\left(T_M^2 + 2\bar{\rho}^2 + 2\varepsilon_M^2 + d_M^2\right)$$
$$+ \vartheta_{z_4}(\mu_{v_a} v_d + d(v_d) + W_4^{*T}\varphi_4(\chi_4) + \varepsilon_4(x_4)) \tag{23}$$

By virtue of Young's inequality, there exists the following inequality as

$$\vartheta_{z_4}(\mu_{v_a} v_d + d(v_d) + \varepsilon_4(x_4)) \le \vartheta_{z_4} v_d + \vartheta_{z_4}^2 + \frac{1}{2}d_M^2 + \frac{1}{2}\varepsilon_M^2 \tag{24}$$

where $d_m$ denotes a positive constant satisfied $|d(v_d)| \le d_M$ . Construct the controller and adaptive law as follows:

$$v_d = -r_4 z_4 - \vartheta_{z_4} - \hat{W}_4^T\varphi_4(\chi_4) \tag{25}$$

$$\dot{\hat{W}}_4 = \vartheta_{z_4}\varphi_4(\chi_4) - d_4 \hat{W}_4 \tag{26}$$

where $r_4$ and $d_4$ are design parameters. Then, inserting the above results into Eq. (23), one has

$$\dot{V}_4 \le -\sum_{i=1}^{4} r_i \vartheta_{z_i} z_i + \sum_{i=2}^{4} d_i \tilde{W}_i^T \hat{W}_i + \frac{1}{2}\left(T_M^2 + 2\bar{\rho}^2 + 2\varepsilon_M^2 + d_M^2\right) \tag{27}$$

In light of the above design processes, a theorem is provided as

Theorem 1: Consider the model of PMSM (2). Suppose the virtual con-trollers and controllers are constructed as Eqs. (7), (13), (19) and (25). Then, the developed adaptive neural control scheme can ensure that all signals are bounded, and all states and control inputs are constrained within predefined intervals.

Proof: Define the overall lyapunov function as $V = V_4$ . With the help of lemma 1 and definitions of $\tilde{W}_i(i = 2,3,4)$ , the derivative of V is derived as

$$\dot{V}_4 \le -\sum_{i=1}^{4} r_i \log\frac{k_{bi}^2}{k_{bi}^2 - z_i^2} - \sum_{i=2}^{4} \frac{d_i}{2}\tilde{W}_i^T\tilde{W}_i + N \tag{28}$$
$$\le aV + N$$

where $N = \frac{1}{2}\left(T_M^2 + 2\bar{\rho}^2 + 2\varepsilon_M^2 + d_M^2\right) + \sum_{i=2}^{4}(d_i W_i^{*T} W_i^*)/2$ , $a = \min\{2r_i, d_i\}$ and the Young's inequality $d_i \tilde{W}_i^T \hat{W}_i \le -\frac{d_i}{2}\tilde{W}_i^T\tilde{W}_i + \frac{d_i}{2}W_i^{*T}W_i^*$ has been used. Meanwhile, the inequality implies $V(t) \le V(t_0) + \frac{N}{a}, \forall t \ge t_0$ . Then, all signals are verified to be bounded. Moreover, there exists $|x_1| \le x_d + |z_1| \le k_{c1}$ , $|x_2| \le z_2| + |\alpha_1| \le k_{b2} + |\alpha_1| \le k_{c2}$ , $|x_3| \le z_3| + |\alpha_2| \le k_{c3}$ and $|x_3| \le z_3| \le k_{b4}$ ,such that all states are constrained within predefined intervals.

## 4. Simulation Results

This section verifies the effectiveness of the developed scheme by several simulation results. The parameters of PMSM is given as follows: $\bar{\sigma} = 5.56$ and $\bar{\gamma} = 14.93$. The reference signal is chosen as $x_d = 0.5\sin(t)$. The constrained constants are given as $k_{b1} = 1, k_{b2} = k_{b3} = k_{b4} = 2, u_{dM} = 1, u_{dm} = -1.5, u_{qM} = 10$ and $u_{qm} = -15$. The other design parameters are designed as follows: $r_1 = 6, r_2 = r_3 = r_4 = 16$ and $d_2 = d_3 = d_4 = 0.3$.

By executing the controllers (19) and (25) for PMSM (1), the simulation results are provided as Figs. 1-3. Fig. 1 displays the curves of rotor position $\Theta$ and reference signal $x_d$, which can be find that the rotor position $\Theta$ is constrained by a positive constant $|\Theta| \le k_{b1} + 0.5 = k_{c1}$. Fig. 2 describes the curves of control signals, which are constrained within an approximate interval. The axis currents are displayed in Fig. 3.

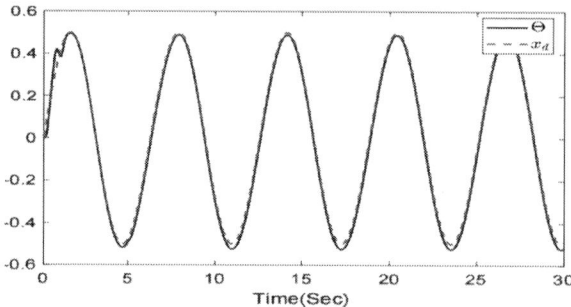

**Fig. 1.** Curves of rotor position $\Theta$ and reference signal $x_d$.

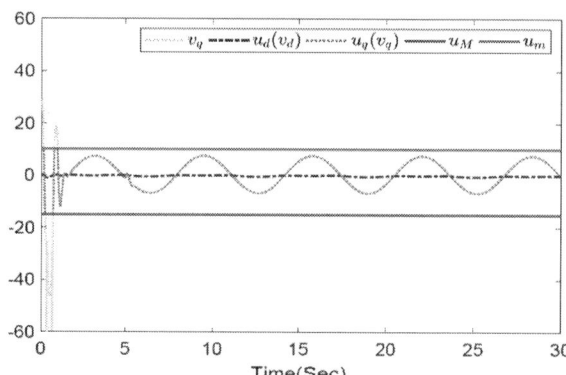

**Fig. 2.** Curves of control signals.

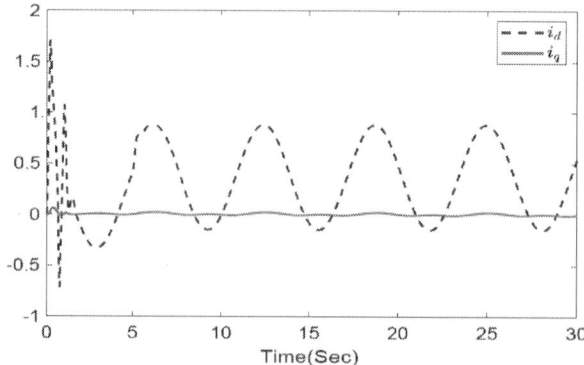

**Fig. 3.** Curves of $d - q$ axis currents..

## 5. Conclusion

This paper investigates the issue of adaptive neural control for chaotic permanent magnet synchronous motor with asymmetric input saturation and state constraints. By employing the first-order slide-mode differentiator, the ``explosion of complexity" issue is avoided. Stability analysis demonstrates that all states and control inputs are constrained within predefined intervals. Finally, simulation results verify the effectiveness of the proposed control scheme. Future work will extend the developed scheme to chaotic permanent magnet synchronous motor with iron losses.

## References

[1] Huixian Liu and Shihua Li. Speed control for pmsm servo system using predictive fuctional control and extended state observer. IEEE Transactions on Industrial Electronics, 59(2):1171 – 1183, 2011.

[2] Jinpeng Yu, Bing Chen, Haisheng Yu, and Junwei Gao. Adaptive fuzzy tracking control for the chaotic permanent magnet synchronous motor drive system via backstepping. Nonlinear Analysis: Real World Applications, 12(1):671 – 681, 2011

[3] Xiaoguang Zhang, Lizhi Sun, Ke Zhao, and Li Sun. Nonlinear speed control for pmsm system using sliding-mode control and disturbance compensation techniques. IEEE transactions on power electronics, 28(3):1358 – 1365, 2012.

[4] Majid Moradi Zirkohi. Command filtering-based adaptive control for chaotic permanent magnet synchronous motors considering practical considerations. ISA transactions, 114:120 – 135, 2021.

# AUTHOR INDEX

Chen, Zhiyuan ........................................................................6

Han, Xiaoyu.........................................................................1

Hu, Yueqiang ......................................................................20

Jiang, Fengguang ................................................................20

Liu, Hongshan ....................................................................41

Liu, Huibo...........................................................................41

Liu, Suyue...........................................................................26

Ma, Limin ............................................................................1

Shang, Jiayi..........................................................................6

Sun, Zhongjiang..................................................................26

Wang, Zhenghua ...................................................................1

Wang, Zhuolin ....................................................................12

Xi, Mingze ..........................................................................36

Xu, Shuangfeng ..................................................................26

Xu, Xianghai ........................................................................6

Yang, Zhiming ...................................................................26

Zhang, Zhipeng ....................................................................6

Zhao, Yanwei .....................................................................41

**Institute of Physics**
Dirac House, Temple Back
Bristol BS1 6BE UK

ISSN: 1742-6588
ISBN 978-1-7138-6775-3

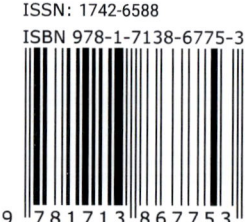